NATO Science for Peace and Security Series C: Environmental Security

NATO Science for Peace and Security Series
Sub-Series C. Environmental Security

Volumes in the Environmental Security Sub-Series deal with scientific and technological aspects of innovative and promising projects related to environmental security, including regional environmental problems and natural and man-made disasters. Floods, droughts, landslides, desertification, earthquakes and industrial accidents are examples of topics covered, as are regional environmental problems related to transboundary water resources or transfrontier air pollution. Other areas covered relate to climate change, environmental chemistry, disaster prevention and forecast, remote sensing and detection, and decision support systems/risk assessment.

Savo Kentera · Zoran Keković · Denis Čaleta
Editors

Building Resilience in a Dynamic Critical Infrastructure Landscape

Cross-Sectoral Challenges

Springer

Editors
Savo Kentera
Atlantic Council of Montenegro
Podgorica, Montenegro

Zoran Keković
Faculty of Security Studies
University of Belgrade
Belgrade, Serbia

Denis Čaleta
Institute for Corporative Security Studies
Ljubljana, Slovenia

ISSN 1874-6519 ISSN 1874-6543 (electronic)
NATO Science for Peace and Security Series C: Environmental Security
ISBN 978-3-032-04382-5 ISBN 978-3-032-04383-2 (eBook)
https://doi.org/10.1007/978-3-032-04383-2

Proceedings of the Advanced Research Workshop on Challenges and Perspectives: Building Resilience in a Dynamic Critical Infrastructure Landscape 11–12 February, 2025, Podgorica, Montenegro

Supporting Team
Ema Kaluđerović
Sanja Grbović
Marina Dabetić

This Springer imprint is published by the registered company Springer Nature Switzerland AG
The registered company address is: Gewerbestrasse 11, 6330 Cham, Switzerland

About This Book

In an age marked by increasing complexity, uncertainty, and disruption, the resilience of critical infrastructure has emerged as both a strategic imperative and a shared responsibility. This volume, part of the NATO Science for Peace and Security Series C: Environmental Security, brings together a spectrum of research, policy insight, and operational experience dedicated to understanding and enhancing the resilience of systems that form the backbone of modern society.

The contributions presented in this book are grounded in the outcomes of the NATO Advanced Research Workshop (ARW) Challenges and Perspectives: *Building Resilience in a Dynamic Critical Infrastructure Landscape*, held on February 11–12, 2025, in Podgorica, Montenegro. Jointly organized by the Atlantic Council of Montenegro and the Center for Risk Analysis and Crisis Management (Belgrade), the workshop convened 48 participants from 13 countries, representing a rich diversity of expertise, institutional affiliations, and regional perspectives. Their shared objective: to analyze vulnerabilities, identify cross-sectoral dependencies, and propose actionable strategies for the protection of critical infrastructure in an evolving threat environment.

As the contributions in this volume illustrate, resilience is not merely a technical or security challenge—it is an interdisciplinary and multidimensional pursuit that demands collaboration across borders, sectors, and domains. From artificial intelligence and biomedical security to road safety and underwater infrastructure, the academic and national perspectives captured here reflect both the breadth and the urgency of the issues at hand.

This book is divided into two thematic parts. The first, Academic Forward, presents theoretical and applied insights into risk management, technological vulnerabilities, and strategic foresight. The second, National Forward, focuses on country-level policies, institutional practices, and regional challenges in aligning with broader European and NATO resilience objectives. Together, these chapters offer a comprehensive view of the critical infrastructure landscape and provide a foundation for further research, policy development, and international cooperation.

We extend our sincere gratitude to all contributors, participants, and supporting institutions whose expertise and engagement shaped this publication. It is our hope that this volume will serve not only as a record of shared knowledge but as a catalyst for continued dialogue, innovation, and preparedness in the face of an increasingly dynamic and interconnected global environment.

Savo Kentera
Zoran Keković
Denis Čaleta

Contents

Part II National Forward

Part I
Academic Forward

Chapter 1
Managing Emerging Risks in Critical Infrastructure—Thinking in Tetris

Zoran Keković, Nebojša Bačanin Džakula, Marina Dabetić, Vladimir Ninković, and Marko Gavrilović

Abstract Managing emerging risks in critical infrastructure requires dynamic, adaptive frameworks capable of capturing complexity, uncertainty, and time-sensitive decision-making. This paper introduces a novel conceptual model based on the game Tetris to illustrate how emerging risks—such as cyber threats, climate impacts, and cascading system failures—can be visualized and managed more effectively. In this analogy, each Tetris piece represents a new risk or disruption entering the system, while the game board mirrors the existing operational landscape with unresolved vulnerabilities. The act of rotating and placing pieces symbolizes organizational risk responses and mitigation strategies, which must be timely, context-aware, and integrated to avoid escalation. As the game progresses and pieces fall faster, the analogy emphasizes how the accelerating pace of risks demands agile foresight, continuous monitoring, and systemic resilience. This approach offers both a pedagogical and strategic tool for understanding risk accumulation, prioritization, and intervention timing within complex critical infrastructure environments.

Keywords Emerging risks · Wicked problems · Risk management · Critical infrastructure · Resilience · Tetris-like thinking

---

Z. Keković (✉) · V. Ninković
Faculty of Security Studies, University of Belgrade, Belgrade, Serbia
e-mail: zkekovic@fb.bg.ac.rs

V. Ninković
e-mail: vladamcc@gmail.com

N. B. Džakula
Singidunum University, Belgrade, Serbia
e-mail: nbacanin@singidunum.ac.rs

M. Dabetić
Department of Physical Chemistry, Vinča Institute of Nuclear Sciences—National Institute of the Republic of Serbia, University of Belgrade, Belgrade, Serbia
e-mail: fmarina@vin.bg.ac.rs

M. Gavrilović
Astronomical Observatory, Belgrade, Serbia
e-mail: mg.uni100@gmail.com

S. Kentera et al. (eds.), *Building Resilience in a Dynamic Critical Infrastructure Landscape*, NATO Science for Peace and Security Series C: Environmental Security,
https://doi.org/10.1007/978-3-032-04383-2_1

1.1 Introduction

Critical infrastructure systems today face a turbulent risk landscape characterized by high complexity and uncertainty. The convergence of digital and physical domains, climate change, and global interconnectivity means that infrastructure operators must contend with volatile, uncertain, complex, ambiguous, and disruptive ("VUCAD") conditions on a regular basis [1]. In such an environment, traditional risk management approaches—often linear and checklist-driven—struggle to cope with the pace and interdependence of emerging threats. Complex risk scenarios in critical infrastructures frequently manifest as wicked problems, wherein any attempted solution can spawn new issues, and no solution is clearly right or wrong [1]. These wicked risk problems defy simple risk matrix approaches, as they have numerous stakeholders, interconnected causes, and evolving dynamics that do not lend themselves to one-off solutions. Consequently, the limitations of conventional frameworks (such as static risk registers or compliance-focused standards) become apparent when facing novel or fast-evolving crises.

Traditional risk management standards like ISO 31000 provide general guidelines for identifying and treating risks, but they were conceived largely of well-understood risks in relatively stable contexts. Advanced complexity and severe unpredictability pose a direct challenge to such frameworks, which were not designed for "unknown unknowns" or rapid cascades of events. Recent research in enterprise risk management highlights that conventional processes focus on compliance and siloed risk assessment, failing to account for nonlinear interactions or cascading failures [2]. In critical infrastructures, this gap can be fatal: seemingly minor issues can escalate through tightly coupled systems, outpacing the ability of a slow, step-by-step risk process to respond. Indeed, infrastructure risk experts note that infrastructures are shifting from merely complicated to truly complex domains, requiring fundamentally new approaches that emphasize agility and adaptation over optimization [3]. Managing risk in these systems is less about eliminating uncertainties—which is often impossible—and more about building capacities to rapidly adapt and respond as situations unfold.

To address these challenges, scholars and practitioners are calling for innovative, adaptive frameworks that can handle the wicked nature of emerging infrastructure risks [1, 2]. Rather than relying solely on static analyses, there is a need for approaches that embrace continual learning, scenario exploration, and real-time decision support. In this paper, we introduce a novel conceptual tool to meet this need: a Tetris analogy for risk management. The popular game Tetris is used as a metaphor and simulation framework for understanding and managing emerging risks in critical infrastructure. We propose that "thinking in Tetris"—envisioning risk management as analogous to the gameplay of Tetris—offers a fresh way to visualize and practice responding to complex, dynamic risk scenarios. Tetris is a fast-paced puzzle in which pieces (tetrominoes) of various shapes fall continuously and must be arranged under time pressure. This paper explores how the key elements of Tetris (falling pieces, the game board, line clearing, etc.) can symbolically map to risk components (emerging threats, the operational environment, risk mitigation actions, etc.), and how the cognitive strategies used in Tetris gameplay parallel those needed for agile risk management.

We hypothesize that the Tetris analogy provides both pedagogical advantages (making training for wicked scenarios more engaging and intuitive) and strategic insights (highlighting the importance of timing, pattern recognition, and adaptation in risk response). Through this framework, critical infrastructure managers could simulate complex cascading crises in a simplified environment, observe the outcomes of delayed versus timely actions, and iteratively improve their strategies in a low-stakes setting. The remainder of this paper is structured as follows. In Sect. 1.2, we review the background on emerging risks, critical infrastructure resilience, the shortcomings of existing standards (like ISO 31000) in addressing wicked problems, and related work in systems thinking and gamification for risk visualization (including the new ISO 31050 guidance). Section 1.3 introduces the Tetris analogy in depth, mapping game mechanics to risk management concepts and discussing the cognitive parallels. Section 1.4 presents our methodology, including a four-layer conceptual model for the Tetris-inspired risk simulation (covering human cognitive requirements, performance metrics, simulation logic, and AI decision support), along with a sample implementation scenario. In Sect. 1.5, we apply the framework to a representative wicked scenario in a cyber-physical critical infrastructure disruption, illustrating how the analogy can capture cascading effects and the value of rapid decision-making. Section 1.6 evaluates the strengths and weaknesses of the approach, comparing its usability to traditional tools and discussing integration into current risk practices. Finally, Sect. 1.7 concludes and outlines future work, including potential development of training tools and further research to refine the Tetris-based risk management approach.

1.2 Background and Related Work

1.2.1 Emerging Risks and Wicked Problems in Critical Infrastructure

Emerging risks are newly developing or changing risks that are difficult to quantify or manage due to their novelty and uncertainty. According to the recently released ISO 31050 standard (Risk management—Guidance for managing emerging risks to enhance resilience), emerging risks are characterized by their newness, a lack of historical data, and insufficient knowledge for decision-making [4]. These are risks that may have been previously unknown, or known but not well understood, and they often come with the potential for very large impacts (threats or opportunities) if they materialize [4]. Unlike well-known risks, emerging risks might not have clear precedents or established mitigation measures. For example, the rise of novel technology, a new type of cyber threat, or an unprecedented pandemic can introduce hazards that evolve quickly and unpredictably.

Researchers and risk bodies have identified several key characteristics of emerging risks that distinguish them from routine issues. Emerging risks tend to be ambiguous (the risk is hard to even define or recognize clearly), chaotic (the situation changes rapidly and nonlinearly), and complex (multiple factors and systems are affected simultaneously) [5]. They often have a shifting time horizon—what once seemed a distant concern can suddenly become immediate 10—and they are marked by deep uncertainty (limited knowledge on probabilities or outcomes) and uncontrollability (often originating outside the organization's direct control, forcing adaptation rather than simple control) [5]. In addition, emerging risks are frequently volatile, meaning their state or severity can change drastically in a short time frame [5]. A vivid illustration was the COVID-19 crisis: experts knew a pandemic was possible, but before it struck it was ambiguous and seemingly distant; once it emerged, it evolved chaotically and required organizations to continually adapt as new information arose [5]. Because emerging risks break the mold of what organizations are used to, they often do not fit neatly into traditional risk assessment templates.

Crucially, many emerging risks in the context of critical infrastructure manifest as wicked problems. The term "wicked problem" (coined by Rittel and Webber in the context of social planning) refers to a problem that is ill-defined, has innumerable causes, no single optimal solution, and any attempt to solve it can lead to new problems. In risk management terms, wicked problems mean that even our solutions have consequences, and we often can at best achieve a satisfactory outcome rather than a clear "fix." A concise definition is: wicked problems are those for which any resolution "only generates further issues," and where solutions are not simply true/false or good/bad, but are better seen as temporary, best-effort interventions based on current information [1]. Critical infrastructure risks are wicked in that they involve complex socio-technical systems, conflicting stakeholder objectives, and evolving threat landscapes. For instance, how to secure the electric grid against cyber threats while upgrading aging equipment and integrating renewable energy is a wicked problem: addressing one aspect (e.g., tighter IT security) might create new issues (such as reduced interoperability or higher costs), and the situation is continually changing with technological advances and threat actor tactics.

Given these attributes, traditional risk management approaches face serious limitations in dealing with emerging and wicked risks. ISO 31000 and similar frameworks outline a risk process (identify, analyze, evaluate, treat, monitor) that works well for static or well-understood risks, but they offer little guidance on how to scan the horizon for unknown threats or adapt on the fly to rapidly evolving scenarios. Often, emerging risks do not announce themselves through clear data or past loss statistics, so a conventional risk register might simply miss them or underestimate them. Moreover, the tendency in traditional risk management to seek definitive estimates of likelihood and impact is frustrated by the profound uncertainty of emerging issues. Indeed, the existence of phenomena like "black swans" (extreme unexpected events) and "unknown unknowns" has led some to argue that standard risk tools are insufficient in today's world of systemic volatility. Studies have noted that classic risk management tends to be reactive and siloed, focusing on compliance and ticking boxes, whereas emerging risk governance demands proactiveness and integration [2].

Recognizing these gaps, the new ISO 31050 guidance was developed to extend the ISO 31000 framework specifically for emerging risks [4]. ISO 31050's primary contribution is a risk intelligence cycle that is iterative and adaptive: it emphasizes continually scanning the external environment for weak signals of change, and then internally analyzing how those changes could impact the organization's objectives [4]. In essence, it formalizes a system of early warning and continuous learning. The cycle consists of two interconnected loops—an external cycle of monitoring the context (technological trends, geopolitical shifts, etc.) for signs of emerging issues, and an internal cycle that builds understanding and assesses the significance of those issues for the organization [4]. By doing so, ISO 31050 aims to help organizations become more anticipative and agile in their risk management. It provides guidance on developing new competencies, metrics, and processes for emerging risk and resilience [6], complementing the more static risk assessment processes with forward-looking insight. This paper aligns with that philosophy: our Tetris-inspired framework can be seen as one imaginative way to cultivate the "risk intelligence" that ISO 31050 calls for, by training decision-makers in a dynamic simulation of emerging risk scenarios.

1.2.2 Critical Infrastructure Resilience and Systemic Interdependencies

Critical infrastructures—such as energy grids, water supply, transportation networks, communications, and healthcare systems—form the backbone of modern society. They are highly complex systems of systems, wherein interdependence abound: the functioning of one infrastructure often relies on others. For example, an electric power grid depends on telecommunications for control signals, while telecom networks depend on power for their equipment; transportation systems need electricity and fuel, which in turn depend on transportation for delivery, and so on. Because of these tight couplings, failures can cascade in unexpected ways across sectors. It has been observed that most infrastructure systems today are interconnected, whether through physical linkages (e.g. shared power, fuel, or water) or through digital means (data, control systems) [7]. As a result, an impairment or disruption in one area (a "trigger" event) can affect multiple other industries or regions, causing an impact far beyond the original point of failure [7]. For instance, a localized power substation failure might lead not only to a blackout but also knock out cellular networks and disrupt water treatment pumps in the region, escalating a single event into a multi-sector crisis.

The resilience of critical infrastructure refers to its ability to withstand, adapt to, and recover from such disruptions. Traditional risk management is often aimed at preventing failures through redundancy and hardening of individual components.

However, in the face of emerging risks and complex interdependence, resilience thinking has shifted toward ensuring systems can adapt and continue operating even when surprises occur. Experts like Chester and Allenby note that infrastructure management is increasingly moving into a domain of complexity where agility and flexibility are paramount [3]. Long-term trends such as climate change, urbanization, and cybersecurity threats are "pushing infrastructures further away from the complicated domain (manageable by optimization) to the domain of complexity," which demands new approaches [3]. In practice, this means infrastructure managers must focus not only on risk prevention but also on response and recovery speed, cross-sector coordination, and learning from near-misses to continuously improve. A resilient critical infrastructure system is one that can take a hit (or multiple simultaneous hits) and still avoid catastrophic collapse by reconfiguring itself or rapidly deploying backup solutions.

One challenge is that complex system behavior can confound linear planning. Infrastructure networks can exhibit nonlinear failure patterns—small triggers might cause outsized effects if they hit vulnerable points, while larger shocks might be absorbed if the system adapts quickly. Therefore, understanding and modeling infrastructure interdependencies is crucial for resilience. As highlighted in a recent CRO Forum report, to properly address business continuity and resilience, one must account for "increasing interdependencies and interconnectedness" and the fast pace of change in infrastructure sectors; doing so often requires sophisticated analysis and simulation tools given the sheer complexity involved [7]. Traditional risk assessments may struggle here, because they often look at risks in isolation or within the boundary of a single organization. By contrast, resilience analysis must consider network effects and second- or third-order consequences of an event.

The limitations of standards like ISO 31000 become evident in this context: while they encourage considering context, they do not explicitly mandate examining cross-infrastructure ripple effects or the kind of systemic risk that critical infrastructures face. Furthermore, resilience is about more than identifying risks about managing the unexpected. This is where adaptive, system-thinking approaches come in (discussed in the next subsection) and where innovative tools like the Tetris analogy may contribute. The gamified approach we propose can illustrate interdependencies by showing how one "piece" can cause trouble if not properly managed, affecting the whole "board" (system state). It also emphasizes the importance of time—a delay in responding to one issue can mean that subsequent issues compound the situation, much as a delay in placing a Tetris piece optimally can complicate the entire stack of blocks.

In summary, critical infrastructure risk management is evolving from a deterministic, siloed practice to a more holistic and agile discipline. Resilience-building efforts focus on maintaining operation through disruptions, which requires anticipation, rapid response, and coordination across sectors. However, achieving this is nontrivial due to the wicked nature of the problems: there is often no single authority or simple fix for system-wide vulnerabilities, and interventions can have unintended effects. This recognition has led to the exploration of new methodologies, including those drawn from systems thinking and complex adaptive systems theory, to supplement classical risk frameworks.

1.2.3 Systems Thinking and Adaptive Approaches in Risk Management

Systems thinking is an approach that views problems in terms of the broader system of interacting components, rather than isolating each component. In risk management, a systems thinking perspective means acknowledging interconnections, feedback loops, and the possibility that a change in one part of the system can have cascading or feedback effects on another. This perspective is essential for dealing with the complex adaptive systems that characterize critical infrastructure and enterprise environments. A complex adaptive system (CAS) is one that consists of many agents interacting in nonlinear ways, adapting to each other's behavior. Business and infrastructure systems can be seen as CAS, since they adapt to changes (e.g., operators respond to incidents, adversaries adjust tactics, markets shift) and thus the risk landscape is always in flux.

Recent thought leadership in risk management has argued for embracing a CAS mindset. Sheth and Sinfield (2024) note that there is a push for a "shift of mindset from mitigating risk to building resilience by honing a CAS view", which means instead of trying to predict and prevent every risk, organizations focus on being robust and flexible enough to handle whatever comes [2]. This entails contextualizing risks within the whole system—understanding how different risks interrelate—and accepting that some level of uncertainty will always persist. Rather than one optimal solution, the goal is continuous adaptation (often termed adaptive management). Adaptive management in this context is an iterative process of decision-making where strategies are adjusted in light of new information or outcomes (essentially learning by doing, under uncertainty). Such approaches have been successful in fields like ecology and are increasingly seen as relevant to infrastructure governance [3]. For example, instead of a fixed plan for disaster response, an adaptive approach would use drills and simulations to learn how the system reacts, update response plans accordingly, and incorporate flexible protocols that can change during an event as needed.

Key principles emerging from the adaptive, systems-thinking paradigm include: (i) promoting a shared understanding among stakeholders of how the system works and what its limits are (so that decisions are informed by a common picture of reality), (ii) treating infrastructures and organizations as systems with dynamic behavior rather than static assets (which implies monitoring indicators of stress and performance continuously), (iii) emphasizing experimentation and scenario exploration over rigid one-time analysis (e.g., using simulations, "what-if" games, and pilot projects to explore system responses), and (iv) fostering a culture of flexibility (rewarding creativity and "satisficing" solutions that are good enough under the circumstances, rather than insisting on theoretical optimums) [3]. All these competencies are aimed at handling a world where change is rapid and objectives can be fragmented or evolving.

When it comes specifically to risk management frameworks, applying systems thinking may involve tools like bow-tie analyses for interdependencies, network models of risk (e.g., Bayesian networks or influence diagrams capturing causal chains), and continuous scanning/forecasting practices (as in ISO 31050's iterative cycles). Our work in this paper resonates with these ideas: the Tetris analogy essentially forces a holistic view (the player must watch the entire board, not just one piece) and encourages on-the-fly adaptation to whatever piece comes next. In a way, it operationalizes systems thinking in a simplified game environment—multiple pieces (risks) interplay on the board and the player must manage the system state (the configuration of blocks) rather than any single block in isolation.

Another aspect to highlight is that time delays and feedback are critical in systems thinking. Many risk failures in infrastructure happen not just from one big shock, but from a accumulation of small issues and delays in responding (feedback loops that amplify a problem). Classic examples include the 2003 Northeast blackout in the US, where slow response to an initial grid malfunction allowed it to propagate, or financial crises where delays in policy response worsen the outcome. Systems thinking teaches that understanding these dynamic behaviors (like how a delay in risk response can lead to feedback that escalates the situation) is as important as understanding the individual components. We will see in the Tetris framework that "delay" in placing a piece has an easily visualized consequence (the stack grows higher, leaving less room for future pieces), which serves as a metaphor for delayed action in real risk management.

In summary, the shift toward systems thinking and adaptive methods in risk management literature and standards (including ISO 31050) underpins the rationale for our approach. We aim to complement this body of work by providing a concrete, gamified model that can help practitioners and analysts internalize these concepts through active learning.

1.2.4 Gamification and Visualization in Risk and Crisis Management

In recent years, gamification—the use of game elements in non-game contexts—and serious games—games designed for education or training rather than pure entertainment—have gained traction in the field of disaster and risk management. The premise is that interactive simulation can engage participants more deeply and help convey complex scenarios in a more intuitive way than traditional lectures or reports. A number of serious games and simulations have been developed to address topics like emergency response, crisis coordination, and risk awareness for both professionals and the general public. For example, disaster management agencies have used tabletop simulation games to train for earthquake response, and humanitarian organizations have experimented with scenario games for pandemic preparedness.

A review by Solinska-Nowak et al. [8] surveyed 45 serious games related to disaster risk management and found that these games serve as valuable educational and engagement tools for a range of stakeholders—from community members to policy-makers. By immersing players in a risk scenario, games can improve risk perception, demonstrate the consequences of decisions, and teach the value of preparedness in a vivid manner. Unlike passive reading, an interactive simulation forces players to make choices and witness outcomes, which can lead to stronger retention of lessons learned. Serious games have been used to model scenarios like flooding, city evacuation, and resource allocation under disaster conditions, giving players a sandbox to experiment with strategies. Importantly, such games allow failure to be a learning mechanism: players can see what happens if they ignore a warning or allocate resources poorly, without real-world damage, thereby learning the importance of timely and effective actions.

Visualization is a key benefit of gamification. Complex risk data or abstract concepts can be represented graphically in a game, making them easier to grasp. For instance, a rising meter or a growing cluster of problem icons on a map in a game can effectively communicate the escalating pressure in a crisis better than a spreadsheet of numbers. In our Tetris analogy, the visual of blocks piling up is an immediate representation of risk accumulation that anyone can understand. This is in line with efforts in risk communication that seek to use visuals and interactive media to make risk information accessible. Gamified simulations can thus bridge the gap between technical risk analysis and human intuition.

There is also evidence that gamified training can enhance decision-making skills under stress. In a press release for a public health crisis game, experts noted that "serious games can help experts explore risk and response to situations that could arise" in scenarios like pandemics and climate crises [9]. By placing players in a time-pressured, information-limited environment (much like a real emergency), games cultivate the ability to quickly interpret data and weigh options. For example, one game might require players (as public health officials) to decide on allocating vaccines with incomplete information about an outbreak, thereby teaching them how to prioritize under uncertainty. Participants of such simulations often report a better understanding of the challenges of coordination and the importance of communication, having lived through a lifelike scenario in the game. Indeed, advanced simulation platforms (like those developed by CAPTRS, as cited in the press release) explicitly aim to "equip decision makers with the intuition and systems needed to combat future ... threats" by exposing them to realistic, unexpected challenges in a game environment [9].

For critical infrastructure, which involves technical systems and human operators, serious games could be particularly useful to train coordination between different infrastructure sectors or between operators and management. Some initiatives have looked at gamifying cybersecurity training for control system engineers or using virtual reality to simulate utility repair scenarios post-disaster. These efforts echo into a common theme: experiential learning is powerful. By 'playing through' a scenario, individuals and teams can uncover pitfalls in their thinking and practice adaptive responses, which is much harder to achieve through traditional drills or meetings.

It's worth noting that while many serious games exist for disaster risk management, using a classical arcade game analogy (like Tetris) for risk management is an unconventional approach. However, the idea falls within the broader trend of using familiar game frameworks to simplify and communicate complex ideas. In our case, Tetris provides a minimalist, universally understood model of a dynamic challenge. By mapping our risk problem onto Tetris, we aim to leverage the game's intuitive visuals and engaging gameplay to demonstrate principles of risk accumulation, mitigation, and strategy.

The next section builds on this foundation by detailing how exactly the Tetris analogy is constructed as a risk management framework. We will map elements of the game to elements of risk systems and elaborate on why this analogy holds cognitive merit for training and analysis.

1.3 The Tetris Analogy as a Risk Management Framework

In this section, we introduce the Tetris analogy in detail and explain how playing Tetris can serve as a metaphorical framework for managing emerging risks in critical infrastructure. The core idea is to create a conceptual mapping between the components and dynamics of the Tetris game and the components and dynamics of a risk management process dealing with continuous, emerging threats. We first outline the mapping of Tetris elements to risk elements, then discuss the cognitive and strategic parallels between the game and real-world risk decision-making, and finally highlight the visual and pedagogical benefits of this analogy.

1.3.1 Mapping Tetris Elements to Risk Components

Tetris is a tile-matching puzzle game where geometric pieces (tetrominoes) fall from the top of a vertical board, and the player must move and rotate them to form complete horizontal lines, which clear out when completed. If the pieces stack up to the top of the board without being cleared, the game is over (failure). We can draw an analogy between these game mechanics and risk management concepts as follows:

- Falling pieces as emerging risks: Each tetromino piece represents an emerging risk or incident that arrives and must be dealt with. Just as pieces appear at random, risks in real life often arrive unexpectedly or with little warning. The different shapes of tetrominoes can be seen as different types of risks, each with its own characteristics and challenges for placement. For example, a long bar piece might symbolize a risk that can cause a broad impact but is easier to place if anticipated (it can clear a line effectively), whereas a complex zig-zag shape might represent a complicated risk that is harder to fit into the existing system (it might leave gaps if not perfectly placed). The continuous stream of falling pieces mimics the continuous flow of new threats or changes in context that an organization faces—one risk after another, sometimes slowly, sometimes in rapid succession.

- The Tetris board as the operational environment: The grid or board onto which pieces fall represents the organization's system or operating environment. This could be the infrastructure network itself or the state of all ongoing organizational activities. The board's width and height correspond to the capacity or tolerance of the system to absorb risk. There is only a finite amount of space to stack pieces, analogous to the limited risk capacity an organization has (for instance, limited resources, limited redundancy, etc.). A piece landing and staying on the board without being cleared is akin to a risk that has materialized or is present in the system—essentially adding to the "load" of issues that the organization is currently handling. The existing stack of pieces at any given time can be thought of as the accumulated risk exposure or unresolved issues. A flat, low stack means the environment is relatively under control (no looming crises), whereas a jagged, tall stack signifies that the system is under stress with many unresolved risk elements piling up.
- Player actions as risk responses: In Tetris, the player can move a piece left or right, rotate it, or accelerate its drop. These actions correspond to risk response actions taken by risk managers. Moving and rotating the piece is like formulating a response or mitigation strategy for the incoming risk—figuring out how to fit this new problem into the current operations with minimal disruption. For example, rotating a piece to fit into a gap is analogous to finding the right contingency plan or resource allocation to address a new threat so that it integrates well and "fills a gap" in defenses. Dropping the piece quickly could be likened to an emergency response (fast-tracking the action because waiting is dangerous), whereas maneuvering it slowly is akin to carefully considering options when time allows. The skill of the player in placing pieces corresponds to the effectiveness of a risk manager's decisions. A well-placed piece (risk handled properly) neatly slots into the existing stack, possibly even resolving some gaps, whereas a poorly placed piece (mismanaged risk) creates awkward overhangs or holes—in risk terms, it could introduce new vulnerabilities or leave issues unresolved.
- Completed lines as mitigated/resolved risks: When a horizontal line is filled in Tetris, it clears from the board, and the player scores points. In our analogy, a cleared line represents a successful risk mitigation or resolution. That means the actions taken by the organization managed to address enough aspects of the risk (and possibly related risks) to effectively remove that threat from the active "stack" of problems. For instance, if multiple small risk events accumulate (like blocks in a line) and the manager implements a solution that neutralizes their combined impact, it's like clearing those issues out—the system returns to a safer state, and capacity is freed up. Cleared lines could also represent learning and improvement: each time the organization fully resolves a set of issues, it could be considered to have gained a reward (points) in terms of experience and resilience.

- Gaps and holes as unresolved vulnerabilities: In Tetris, if pieces are not perfectly placed, gaps can be left beneath the stack which cannot be filled until the right shape comes along, if at all. These holes are analogous to unresolved issues or vulnerabilities that remain after a risk response. For example, an organization might hastily patch a software vulnerability (drop a piece) but in doing so, bypass some best practices, leaving a "hole" that represents a latent vulnerability or a follow-on risk (perhaps a misconfiguration) that persists. Those holes make future risk placements more difficult—similarly, unresolved issues constrain how well you can respond to new risks, because they create a suboptimal baseline state.
- The stack reaching the top as systemic failure: If the pieces stack to the top of the board, the game is over. In our analogy, this corresponds to a critical failure or crisis where the system's risk capacity is exceeded. Essentially, the organization can no longer manage the incoming risks—they are overwhelmed. This could translate to scenarios like a cascading infrastructure blackout, a financial institution collapse, or any situation where too many problems happen at once without sufficient mitigation, leading to total system breakdown. The goal in both Tetris and risk management is to avoid this outcome by keeping the stack of risks down through effective placements (mitigations).
- Increasing speed and complexity: As Tetris progresses, pieces typically fall faster, and the game becomes more challenging. This mirrors how emerging risk situations often escalate over time—threats can compound, and events can accelerate. For instance, a brewing crisis may start slow, but then multiple issues hit in quick succession (like Tetris pieces raining down faster and faster). A risk management analogy is the concept of a crisis snowballing if early risks are not dealt with promptly. The increasing difficulty in Tetris forces the player to improve their reaction time and strategy; similarly, an intensifying risk scenario tests the organization's adaptive capacity and may require faster decision cycles and more automated responses (or additional support, like bringing in an "AI player" as discussed later).

The above mappings form the foundation of the Tetris risk framework. In summary, the falling pieces represent a stream of emerging risks; the board is the system they threaten; the player's maneuvering of pieces are the risk mitigations; cleared lines are resolved risks (improved system state); and an overflow of pieces is system failure.

1.3.2 Cognitive and Strategic Parallels

The Tetris analogy is not only structurally convenient; it also carries meaningful cognitive and strategic parallels to real-world risk management. Playing Tetris engages certain mental processes and decision- making strategies that are strikingly similar to those required of risk managers dealing with fast-moving crises:

- Situational awareness and big-picture thinking: A skilled Tetris player maintains awareness of the entire board configuration while also tracking the current falling piece (and typically the next few pieces in the queue, in modern Tetris versions). This is directly analogous to a risk manager maintaining situational awareness of the overall operational context while simultaneously handling a specific issue. Good risk management requires seeing the big picture (how various risks and factors interrelate) and not just focusing tunnel-vision on one problem. Focusing on a single threat without regard to the broader situation can lead to suboptimal outcomes or unintended side effects.

- Prioritization and resource allocation: In the game, players often have to decide which area to fill or which line to complete, sometimes leaving certain gaps temporarily in order to address more urgent ones. This mirrors prioritization under pressure in risk management. For example, if two issues arise at once, a manager might have to choose which one to tackle first. The strategy is about making trade-offs—very much what crisis managers do when they, say, allocate limited fire crews to the most critical areas of a spreading wildfire, knowing that some smaller fires might burn a bit longer.
- Time pressure and decision speed: Tetris is famously a game of speed. This recreates the time-critical decision-making aspect of emergency risk management. In a spreading outage or cyber-attack, responses often have to be quick to be effective; delays of even minutes can drastically change outcomes. The game enforces a habit of making fast decisions based on incomplete information, which is exactly the challenge in emerging risk scenarios where one must act on partial data. The cognitive skill here is not just speed, but knowing when you can take a moment to think versus when you must act immediately (in Tetris, sometimes you can "soft drop" a piece slowly to buy time, analogous to taking a moment when the situation allows; other times you must "hard drop" because another piece is coming).
- Pattern recognition and anticipation: Expert Tetris players develop the ability to recognize patterns in the blocks and anticipate the best placements quickly. They might see that a T- shaped piece would perfectly fill a certain gap two moves ahead, and prepare for it. In risk management, experienced practitioners similarly recognize patterns in evolving situations—for instance, the telltale signs of an impending system failure or the typical cascade path a certain event might follow. This foresight is crucial: one tries to anticipate what next risk might emerge if the current one is not handled properly. The Tetris analogy encourages thinking ahead. A risk manager often has forecasts or early warning indicators (e.g., weather radar for an incoming storm) to consider. The analogy thus can incorporate the idea of early warning.
- Managing mistakes and adaptive learning: In any Tetris game, mistakes are inevitable—a piece is mis-rotated or dropped in a less-than-ideal spot, creating a troublesome gap. What distinguishes good players is their ability to recover from mistakes: they adapt their strategy to work around the gap, perhaps by clearing lines elsewhere or waiting for the one piece that can fill it, and they try not to compound the error by panicking. In risk management, organizations also make mistakes (or simply have unavoidable residual risk) and must adapt. A classic example is when a containment effort fails partially—rather than giving up, responders must reorient and still find a way to mitigate the damage. Tetris teaches

resilience in this sense: a misplacement doesn't immediately end the game; you can often compensate with a clever placement of the next pieces. This reflects the principle of graceful degradation in systems—even if something goes wrong, the system should adjust and continue, if possible, not collapse outright. The iterative play of Tetris (start a new game after failure) is akin to iterative learning cycles in risk management (after each incident or exercise, update plans).
- Keeping options open (flexibility): Skilled players often try to keep the stack as flat as possible and avoid creating structures that are hard to fill (for instance, leaving a single-column gap open for the long bar piece, a known strategy, but ensuring not to depend on a piece that might not come in time). A risk manager might, for example, preserve some reserve resources or backup plans rather than using everything on the first problem, in anticipation that another issue might follow. It's an exercise in balancing current risk mitigation with preparedness for the next hit.

Through these parallels, we see that the Tetris analogy is more than a playful comparison—it encapsulates many strategic principles of dynamic risk management: maintain situational awareness, prioritize under pressure, act with urgency when needed, anticipate future events, recover and learn from mistakes, and stay flexible. By engaging in the Tetris framework, risk practitioners can practice these cognitive skills in a low-risk environment and potentially transfer those skills to real situations.

1.3.3 Visual and Pedagogical Advantages of the Analogy

The use of Tetris as a metaphorical framework offers distinct visualization and teaching advantages for understanding emerging risks:

- Intuitive Visualization of Risk Accumulation: One of the hardest things to convey in conventional risk assessments is how risks can accumulate over time and interact. Spreadsheets or risk registers list risks separately with static scores, which doesn't show how a sequence of issues might build up to a crisis. The visual of an increasing stack is a powerful analogy for increasing risk exposure. This visual cue can help stakeholders who are not risk experts (e.g., executives or public officials) grasp the importance of proactive risk mitigation. It answers the question "what happens if we delay action on emerging issues?".
- Immediate Feedback and Learning by Doing: Tetris provides instant feedback for each action—a good placement yields a cleared line or a stable stack, a bad placement creates a gap or contributes to a looming tower. In a pedagogical setting, this immediate cause-and-effect is invaluable. This kind of experiential learning is known to be effective in adult education. By doing, participants often internalize lessons much better than by listening. The game element adds a layer of engagement that keeps people interested— there's a reason Tetris is addictive and captures focus. In a training context, that focus means participants are likely to

be more immersed in the scenario, and thus more likely to absorb the underlying risk management lessons.

- Simplification without trivialization: The Tetris model strips down the complexity of real life to a more manageable, simpler representation (blocks on a grid). Yet, it retains the core challenges (multiple issues, limited space/resources, time pressure, need for strategy). This simplification is useful pedagogically because it allows discussions to center on fundamental principles of risk management without getting lost in domain-specific jargon or highly technical modeling. Participants from different critical infrastructure sectors can all relate to it, because it's not locked to one domain (power, finance, etc.); it's an abstract metaphor, which means the focus can be on cross-cutting skills (decision-making, timing, adaptation) rather than sector-specific details. At the same time, the stakes and parameters can be adjusted to mimic complexity. In educational psychology, such abstraction can help learners transfer knowledge to various contexts, because they learn the pattern rather than a single scenario.
- Engagement and Motivation: The game-like nature of the approach can make training sessions or workshops more engaging. The scoring and leveling aspects of Tetris can introduce a bit of gamified competition ("who can survive the longest without a system failure?"), Engaged participants are more likely to remember and later apply what they learned. Additionally, the Tetris analogy could serve as a memorable mental image. Even long after a training, a manager might recall "Keep the stack low!" as a shorthand for addressing small problems before they accumulate, simply because they recall the game.
- Communication: Analogies are powerful communication tools. Explaining a complex risk concept to non-experts or decision-makers can be challenging, but relating it to a simple game can break down barriers. The Tetris metaphor provides a common reference point. One could imagine even in a real crisis meeting, someone saying "We have to clear some lines or this will reach the top"—a metaphorical way to urge action that everyone immediately understands. Thus, beyond training, the analogy might influence organizational culture, providing a vocabulary for discussing risk in a more relatable way. By making these points visually and metaphorically clear, it supports better cross-disciplinary understanding.

The Tetris framework is not meant to replace detailed risk analysis tools for quantitative assessment or precise scenario modeling; rather, its value lies in conceptual insight and human factors training—helping people think differently about risk and prepare mentally for emergent, high-pressure situations. The next section will describe how we build on this analogy to create a working conceptual model and simulation environment, and how we incorporate human and AI elements to enhance its usefulness.

1.4 Methodology

To operationalize the Tetris analogy for risk management, we propose a four-layer conceptual model that forms the basis of a Tetris-inspired risk simulation environment. The four layers address different aspects necessary for simulating emerging risks and training decision-makers: (1) the Cognitive Layer, capturing human cognitive requirements and situational awareness; (2) the Human Performance Layer, defining how we measure and evaluate the player's (risk manager's) performance; (3) the Simulation Layer, which encompasses the mechanics of the Tetris-like risk environment and the outcomes it generates; and (4) the AI-Support Layer, which integrates artificial intelligence to enhance decision support and scenario complexity. After describing each layer, we present a sample scenario demonstration to illustrate the model in action and then discuss implementation considerations for a prototype system.

The results underline the importance of targeted capacity-building measures, including regular interagency training, improved access to detection and protection resources, and better dissemination of standardized operational procedures. Enhancing institutional coordination and investing in human and technical capacities are essential steps toward a more resilient and responsive CBRN emergency management system.

Future research should continue to explore the role of first responders in various national contexts, with a particular focus on organizational culture, perceived preparedness, and inter-institutional collaboration.

1.4.1 Four-Layer Conceptual Model Overview

Layer 1: Cognitive Layer (Human Decision Requirements). This layer defines the cognitive tasks and information cues available to the human participant (the risk manager "player"). It includes what the human needs to perceive and understand at any given moment of the simulation. Key elements in this layer are: Visual dashboard: analogous to the Tetris game screen, showing the state of the system (board and falling piece). This provides situational awareness. The Cognitive Layer ensures that the participant can see current risk exposures (stack configuration) and the incoming risk (current falling piece, and possibly a preview of the next one if we include that feature as an early warning). Controls and levers: what actions the human can take (move/rotate/drop pieces, or in risk terms, deploy certain mitigations). This includes any available options for responding to a risk, which must be presented clearly to avoid decision paralysis. Cues and alerts: indicators for important information such as a warning if the stack is dangerously high (akin to alarms in real systems when risk levels are critical), or a highlight on certain pieces if needed (for example, color-coding types of risks). Cognitive workload considerations: since emerging risk situations can overload human decision-makers, this layer considers how the game can simulate stress (perhaps through increasing speed or injecting simultaneous pieces that mimic multi-tasking demands) and how to keep the information presentation manageable (avoiding clutter on the screen). Essentially, we structure the simulation to exercise the cognitive skills identified in Sect. 3.2 (situational awareness, pattern recognition, etc.) without exceeding realistic human limits.

Layer 2: Human Performance Layer. This layer deals with capturing and evaluating the human's performance in managing the risks within the simulation. It defines metrics and observation points such as: Response time: how quickly the participant reacts to each new "risk piece" (e.g., time from piece appearance to placement). This can be recorded to assess whether quicker decision-making correlates with better outcomes or if rushing causes mistakes. Decision quality: a qualitative or quantitative measure of how well the piece was placed. For instance, did the placement clear lines (resolve risks) or create gaps (residual issues)? We can assign scores to placements (just as Tetris gives points for line clears) to represent effective risk resolutions. A perfect clear might score high (like fully mitigating a risk), whereas a placement that leaves a hole might score low (partial mitigation, leaving a vulnerability). Workload and errors: track sequences where the player struggles—e.g., rapidly rotating a piece back and forth might indicate indecision or confusion (comparable to analysis paralysis in a real crisis), or dropping in a bad spot is an error. We can measure error rate (misplacements) and how the player recovers. If we had biometric or subjective feedback, we could also gauge stress, but that might be beyond scope. Learning curve: if the game is played in multiple rounds, this layer would track improvement over time, such as reduction in average stack height or longer survival before "game over." This reflects whether the participant is learning better strategies, analogous to training effectiveness in a real-world program. Collaboration metrics (if multiplayer or team-based): Optionally, the framework could allow multiple players (like different roles managing the same board, e.g., one handles cyber risks pieces, another physical—simulating interdisciplinary teamwork). In such cases, performance could include communication quality, number of conflicts or synergies in piece placement, etc. However, initial implementation might focus on single-player for simplicity.

Layer 3: Simulation Layer (Risk-Tetris Engine and Outcomes). This is the core mechanics layer, where the analogy is turned into a working simulation: Piece generator (Risk scenario generator): a module that determines what piece (risk) comes next, and at what timing. This can be a simple random generator (like classic Tetris random sequence) or a more complex scenario script. For example, we could script a scenario where initially pieces come slowly (peacetime operations with few incidents), then an event triggers a flurry (crisis onset with many incidents). The generator could be informed by real data or models if desired (e.g., using Poisson processes for random incident arrivals or based on known threat patterns). Board state updater: this computes the effects of each placement. It must apply the rules (clear any full lines, add score, etc.) and also log the risk state. For instance, clearing a line might correspond to resolving several related risks at once, which could be logged as such. The outcome we get from the simulation at any time includes the current stack height (risk level), number of cleared lines (risks resolved), and maybe some indicators (like "near failure" if height above a threshold). Outcome measures: The simulation layer defines what constitutes "game over" (system failure). Typically, it's piece placement beyond the top. In risk terms, we could equate that to an incident that can't be managed—say the organization is overwhelmed. We can also define intermediate outcomes, like how many risks were handled before failure, which is akin to resilience measure (how much shock the system took before collapsing). Scenario parameters: This layer would allow configuring different "board conditions" to represent different systems. For example, a taller board (more rows) is like an organization with more buffer or redundancy (more risk capacity), whereas a narrower board (less columns) might represent a tightly constrained system with fewer ways to reconfigure resources. Similarly, initial pieces already on the board could simulate an ongoing crisis that's already partially developed when the simulation starts (starting with some stack). Stochastic elements and variations: to keep the training realistic, the simulation can incorporate variability. In Tetris, sequences are random; in real life, crises are unpredictable. Running the simulation multiple times with different random seeds ensures that the participant can't just memorize a sequence—they have to practice true adaptivity. If an AI or instructor is controlling the piece generation, they could also introduce deliberate challenges (e.g., three difficult pieces in a row to simulate a multi-hazard event).

Layer 4: AI-Support Layer. The final layer integrates artificial intelligence as both an aid and a challenge to the human player: AI advisor: On the aid side, an AI agent could analyze the board and suggest optimal moves (risk responses). This could take the form of a "ghost piece" showing where the AI would drop the current piece, or a hint system that highlights potential line-clearing placements. In risk terms, this mimics a decision support system that gives recommendations based on data (for instance, an AI analyzing sensor data might suggest focusing on a certain area of the network). The human can choose to follow or ignore the AI advice, and this interplay can be instructive (e.g., if the AI consistently suggests better moves, the human can learn from it; if the human outperforms the AI in some cases, it can highlight the value of human intuition). AI risk generator/opponent: The AI can also control scenario difficulty dynamically. For example, if the human is doing very well (keeping the stack low), the AI might increase the speed or introduce trickier pieces (akin to an adversary exploiting the fact that the defender is comfortable). This dynamic difficulty adjustment keeps the training challenging and can simulate an intelligent adversary in scenarios like cybersecurity (the AI "adapts" like a hacker escalating attacks). Essentially, the AI can ensure that the scenario tests the boundaries of the human's skill, which is useful for evaluating performance under tougher conditions. Machine learning analysis: Over multiple sessions and many players, the AI layer can accumulate data on decisions and outcomes. Using machine learning, it could potentially identify patterns of common mistakes or successful strategies. This can feed back into refining training. For instance, if analysis shows that players frequently leave a particular configuration of gaps that leads to failure, that scenario can be emphasized in future training or highlighted during debrief ("we notice you often leave a single-column gap on the side, which is risky unless a long piece arrives; consider addressing that earlier."). Autonomous agent benchmarking: We could also develop an autonomous agent to play the game (reinforcement learning has famously been used to train AI to play games like Tetris or chess). This agent can serve as a benchmark for optimal play. In risk terms, that's like having an AI risk management system and comparing human performance to it. Such comparisons could identify where humans excel (maybe creativity in unexpected situations) vs where AIs excel (processing lots of data quickly). This layer thus supports not only the human in the loop but also the study of human-AI teaming for managing risks. In a real critical infrastructure control room, human operators increasingly work with AI-based monitoring systems [9]. Our simulation provides a controlled setting to explore that interaction.

1.4.2 Implementation Considerations

To illustrate how these layers come together, consider a sample scenario using our Tetris-inspired model, focused on a cyber-physical critical infrastructure disruption. This sample scenario, while simplified, shows how the Tetris model can simulate a wicked critical infrastructure risk sequence, capturing the interplay of events and responses over time. Next, it can be considered how one would implement such a simulation and framework in practice, and how it could be used in real training or risk analysis settings. In implementing the above, we plan to start small: a single-player prototype with a basic piece generator and logging, and one or two pre-scripted scenarios, which is out of the scope of this paper. In essence, the scenario and its outcome provide a concrete basis for discussing best practices in an emergency: the importance of early containment of problems, maintaining flexibility, and the dangers of indecision.

Implementing the Tetris-inspired risk simulation involves both software development and careful design of scenarios to ensure realism and relevance. This prototype can then be used in a pilot study with risk management professionals to gather feedback on realism, usability, and learning outcomes. The insights from that will drive further refinement.

1.5 Application to Wicked Scenarios in Critical Infrastructure

In this section, we apply the Tetris-based risk management framework to a complex wicked scenario involving a cyber-physical disruption cascading across interconnected critical infrastructures.

1.5.1 Scenario Description: "Blackout Cascade"

Context: Imagine a national critical infrastructure network encompassing the electric power grid, telecommunications, and transportation control systems. It's a cold winter, and the system is already under moderate strain supplying heating power. This scenario spans multiple sectors and has high uncertainty—a hallmark of wicked problems because multiple independent actors (power companies, telecom providers, transport authorities, government emergency agencies) are involved, no one has complete information, and the situation evolves quickly in surprising ways.

- Initial trigger: A sophisticated cyber-attack is launched against the power grid's control centers (perhaps by a state-sponsored hacker group). This is analogous to the first falling piece in Tetris. The attack introduces false sensor readings and disrupts grid balancing in a region. As a result, a major transmission line unexpectedly trips. The operators see conflicting information—some systems flag overload, others appear normal (reflecting ambiguity).
- Cascading impact: The loss of the transmission line causes power flows to reroute rapidly. Within minutes, neighboring lines overload as well. A cascade of line failures occurs, leading to a partial blackout in several states/provinces. This is now multiple pieces stacked—a large portion of the power grid is down, and the risk "stack" is growing fast.
- Cross-sector effects: The blackout doesn't just mean no electricity; telecom networks in the affected areas begin to fail as cell towers and central offices exhaust their battery backups. Traffic lights go dark, causing gridlock in cities (transportation infrastructure affected). Emergency services are hampered by communication outages. Now we have a multi-faceted crisis: power, communications, and transport all at once. In Tetris terms, the pieces are coming from different angles—it's as if one had a mix of wide and narrow pieces creating an irregular structure. No single agency or simple action can solve this; it requires coordinated responses, another reason it's wicked.
- Managing the crisis (analogy to playing the game): Delays versus agile decisions: In the early moments, if grid operators hesitate (say they doubted whether it was a cyber-attack or a sensor glitch and delayed tripping some lines to isolate the problem), the cascade grows. This is akin to the Tetris player hesitating and letting pieces fall in bad spots. In contrast, if they acted quickly to isolate the affected region (shedding load deliberately, a bit like sacrificing part of the stack to flatten the rest), the cascade might have been contained, preventing the wider blackout. This difference in approach reflects a core lesson: time is critical. A delay of even a few minutes in initiating load-shedding or starting backup generators for telecom can be the difference between a local outage and a regional disaster. Our simulation model demonstrates that—early line clears (mitigations) keep the stack low, whereas delays cause piling.
- Response speed and adaptation: As the blackout cascades, responders must adapt strategy dynamically. Initially, the focus might be on stabilizing the power grid (clearing lines in one color). But as communications fail, priorities shift—maybe

the next "piece" is to get mobile generators to cell towers to restore critical communication links. The crisis managers are essentially playing multiple Tetris boards: one for power restoration, one for telecom, etc., and moves on one board affect the other. If simulation is a single board, one could envision it segmented or color-coded by sector, requiring balance. The lesson here is adaptability: plans made at the beginning of the event might need to change as new information arrives. A rigid plan (like always trying to clear the power issue first) might fail if communications completely die in the meantime. In the analogy, a player who insists on waiting for the perfect piece to fix one side of the board might neglect the other side where pieces are piling up—leading to loss. So, agile response means continuously re-evaluating priorities: sometimes you clear a less-than-perfect line (partial mitigation) to buy breathing room, rather than chasing an ideal solution that might come too late.

- Delays illustrated: To give a concrete number, in the 2003 Northeast US blackout, it's noted that a failure to trim tree contact on a line and then a software alarm failure led to a ~ 1-hour delay in recognizing the cascade, by which time it was uncontrollable. In our game terms, that was like missing opportunities to clear lines early, eventually the pieces stacked beyond recovery. On the other hand, in 2019, a smaller scale incident in the UK power grid was contained within minutes by automated load shedding, preventing a nationwide outage—akin to quickly clearing lines to keep stack low.
- Learning from mistakes: As the scenario unfolds, it's likely mistakes will be made by responders (nobody handles a wicked problem perfectly the first time). Perhaps a region was unnecessarily blacked out due to over-shedding, or a communication to the public was mismanaged causing panic (side effects). These are like the holes left in the Tetris stack. However, those mistakes become learning points for the next round of the crisis or future crises. For instance, after stabilizing the immediate situation, agencies might realize they need better cyber protocols or more robust communications backups. Our simulation encourages this iterative learning: one can replay the scenario and try a different approach, illustrating how outcomes change. In a training, participants often dramatically improve in a second run, which mirrors how real organizations improve after drills or actual events.
- Strategic adaptation and long-term changes: Wicked problems often require not just reacting to immediate events but adapting strategy for the long haul. In our scenario, once the immediate cascade is halted, the broader challenge remains: how to bring systems back up and how to prevent a recurrence. There might be debates about whether to rebuild systems differently (e.g., more microgrids to isolate failures, changes in cyber defense strategies). In Tetris, after a chaotic sequence, a good player might change their strategy—e.g., decide to stack pieces flat and low for a while (focus on stabilization) rather than aggressively clearing (taking big risks). In crisis management, this could equate to switching from an offensive approach (trying to restore everything quickly) to a defensive one (ensuring no further collapse even if it means some services are down longer).

This scenario underscores several key themes that our Tetris framework surfaces: Interconnectedness: The multiple shapes and simultaneous stack issues in the game reflected multiple sectors failing together. It forces the recognition that you can't solve one in isolation easily—just like infrastructure sectors. Urgency versus correctness trade-off: Sometimes you just have to place the piece somewhere acceptable rather than finding a perfect spot, because another piece is coming. Emergency decisions often have to be made with imperfect information—you can't wait for all data or an optimal solution when sirens are blaring. Resource allocation under constraint: The limited board space is akin to limited resources. In the scenario, finite repair crews and backup generators had to be allocated. The game can simulate this by limited space for certain pieces or limiting how often you can "hold" a piece. It trains the idea that you can't save everything; you allocate where it counts most. Delayed effects: The scenario could also incorporate time-delayed risks. For example, a dam might fail hours after the power outage if pumps are off. In Tetris, a delay might be mimicked by, say, a piece that disappears and reappears lower (just a thought) or a slow falling piece that if not managed will eventually hit. This teaches looking ahead to not just what's happening now but what could happen soon (like pieces in the preview). Multi-stakeholder coordination: Our analogy currently has one player, but think of that player as a composite of the whole response team. If they coordinate well internally, the pieces fit; if they argue or act incoherently, pieces collide (like two people trying to move the piece different ways). A possible extension is having multiple players controlling different types of pieces needing to coordinate. The wicked scenario definitely involves multi-agency coordination, which is complicated since each has its own objectives and info. The game could incorporate that by, say, giving two players different information (one sees only certain pieces) and they must communicate to succeed.

By applying the framework to the "Blackout Cascade" scenario, we see how it can encapsulate the essence of wicked critical infrastructure crises: numerous interdependent problems, high stakes, uncertain evolving situations, and the necessity of quick, adaptive, and collaborative decision-making. The Tetris analogy distills these aspects into an experiential form that, while simplified, can drive home lessons about timing, prioritization, and adaptation that are otherwise hard to convey.

1.6 Evaluation and Discussion

1.6.1 Strengths of the Tetris Analogy Approach

- Enhancing Dynamic Thinking: One of the biggest advantages of this approach is that it forces users to think dynamically and adaptively. Traditional risk assessments are often static snapshots (e.g., a risk register is a list at a point in time), whereas our approach continually evolves, more akin to real life. This trains practitioners to continuously scan and respond, aligning with ISO 31050's emphasis on iterative risk cycles [4]. It cultivates an agile mindset: instead of "we did a risk assessment last quarter," it's "we are constantly in a risk assessment, every piece drop is a new assessment." This prepares organizations for unpredictable environments in a way static models cannot.

- Intuitive Communication and Engagement: The Tetris metaphor translates complex risk interactions into a simple visual language of blocks and lines. This is a strength when communicating to stakeholders who may not be versed in risk jargon. For example, an executive might not grasp a heatmap of interdependency vulnerabilities immediately, but they can certainly understand when shown a Tetris simulation that if small problems aren't resolved, the blocks pile up dangerously high. Thus, the approach can help bridge communication gaps between risk analysts and decision-makers. Additionally, the engaging nature of a game means higher participation and interest. In training or workshops, participants are likely to be more involved and attentive during a simulation game than during a slide presentation. This increased engagement can lead to better retention of knowledge.
- Holistic Systems Perspective: The game inherently requires seeing the whole board; you can't succeed by focusing on just one column. This echoes the systems thinking perspective that is needed for wicked problems. It illustrates emergent properties (like how a gap plus a certain piece shape can cause a problem, analogous to two minor risks interacting to create a major issue). Therefore, it's a tool for reinforcing systems thinking in a tangible way. This can complement analytical tools: after using the game, a team might be more cognizant of interdependencies when they go back to fill out their risk registers, for instance.
- Demonstrating Importance of Timing and Sequence: Traditional risk matrices may rank risks by impact and likelihood but miss the importance of when and in what order events occur. Our approach vividly demonstrates how sequence and timing can change everything. This is a strength because many risk methodologies struggle to incorporate time dynamics. The game makes clear that two identical sets of events can lead to totally different outcomes depending on response timing—an insight that can prompt improvements in real processes (like investing in faster detection systems or delineating clear thresholds for action). It also underscores the value of early warning systems and quick decisions, which might justify investments in things like AI monitoring.
- Encouraging Learning and Continuous Improvement: The framework's game nature supports iteration and learning by doing. This is essentially fostering a culture of continuous improvement in risk management. Often, real learning in risk management happens after disasters (through post-incident reviews), but here we can simulate disasters repeatedly, accelerating the learning cycle without real harm. Moreover, the data collected can help identify weaknesses (maybe participants always struggle with a certain type of risk/piece), informing where training or process improvements are needed.
- Integrating Human Factors into Risk Planning: By involving an actual human (or team) making decisions in real-time, the approach accounts for human factors like stress, communication, and decision errors. Many analytical risk tools assume

rational, optimal decision-making or have simplistic decision nodes. Tetris-based simulation acknowledges that humans under duress may make mistakes or brilliant improvisations. It provides a way to observe those behaviors and factor them into planning. For example, if multiple runs show that people consistently overlook a certain type of risk when multitasking, one might implement a real-world procedural check (like always having a second pair of eyes on that risk type during crises). In essence, it can reveal the operational realities that glossy procedures might miss—similar to how pilot flight simulators reveal which emergency procedures are hard to follow in practice.

- Potentially Quantitative Measures for Otherwise Qualitative Concepts: Concepts like resilience, agility, and adaptive capacity are often spoken of qualitatively. The game can provide some quantitative proxies: e.g., "lines cleared" could be a measure of resilience (how much disturbance you handled) and "max stack height reached" a measure of vulnerability (how close to failure you got). While these are game metrics, they might correlate to real metrics (like system reserve margins, etc.). It opens interesting avenues for measuring improvements (e.g., after training, a team's average 'score' improves, implying better teamwork or decision process).

1.6.2 Weaknesses and Limitations

- Oversimplification Risk: By design, the Tetris analogy is a simplification of reality. Critical infrastructure systems and risks have many attributes that cannot be captured in falling blocks—for instance, the geographic distribution of effects, the political and social dimensions of crises, legal/regulatory constraints on actions, etc. One weakness is that stakeholders might misinterpret or oversimplify the real world after using a simple model. There's a danger of the "law of the instrument" ("if all you have is a hammer, everything looks like a nail")—people might start framing all problems in terms of blocks and lines even when that may not be fully appropriate. We must stress that the game is a learning tool, not a literal model of all risk aspects.
- Not a Substitute for Detailed Analysis: The Tetris framework does not provide quantitative risk estimates (like probabilities of failure, expected losses) which are often needed for decision-making (e.g., to justify investments or insurance). It won't output a neat risk register or cost-benefit analysis. Therefore, it cannot replace traditional risk assessment techniques for formal purposes. It's complementary. Organizations still need their ISO 31000 processes, risk matrices, scenario analysis, etc., for documentation and rigorous analysis. Our tool is more about process (how to think and react) rather than analysis outputs. Some traditionalists might see it as a fun gimmick that doesn't produce the hard numbers or reports they need, and thus might undervalue it.
- Validity and Realism Concerns: There might be skepticism about how a game can reflect reality. While we can map many concepts, the lack of one-to-one

realism is a limitation. For example, in reality one cannot "clear away" a risk permanently as neatly as a Tetris line clears; there might be lingering effects. Also, some risk treatments take days or months (vaccinations in a pandemic) which is hard to simulate in a fast game. The time compression and abstraction mean not every important factor is represented (for instance, public communication, or bureaucratic delays due to decision approval chains). If users don't find the scenarios believable, they may not take the lessons seriously. We might consider injecting narrative or debriefs to explicitly connect game actions to real actions, to help transfer the insights.

- Potential to Encourage Gamey Behavior Over Reality: Players might learn to "game the game"—exploiting quirks of the simulation that wouldn't translate to real life. For instance, maybe in the game, one strategy is to always wait for the long bar to clear four lines at once (a known Tetris strategy), but in real risk management, waiting for a big solution could be dangerous. If the game's scoring incentivizes certain actions (like clearing multiple lines at once yields more points), participants might chase those game rewards in ways that conflict with real-world optimal strategies. To counter this, game design must align incentives with real best practices as much as possible (perhaps balancing the scoring).
- User Stress and Accessibility: Not everyone enjoys or excels at fast-paced games. Some individuals may freeze under the time pressure or get frustrated. If a training participant feels overwhelmed by the game, they might disengage or feel demoralized ("I keep failing, so I must be a bad risk manager," which might not be true—they might just be inexperienced with video games). We have to ensure the tool is used in a supportive learning context, not as a high-stakes evaluation unless properly calibrated. Additionally, consider physical limitations or accessibility: not everyone can handle flashy moving graphics or quick hand-eye coordination tasks (some could have disabilities or simply lack gaming experience). We might need alternative modes (like a slower turn-based mode) to make it inclusive.
- Resource Requirements to Implement and Maintain: Building and integrating such a simulation is not trivial. It requires interdisciplinary effort (risk experts, game developers, cognitive scientists). Maintaining its relevance means updating scenarios as new types of risks emerge (imagine 10 years ago, "social media misinformation" wouldn't be in our model). In terms of cost-benefit, one might question: couldn't similar benefits be achieved with a well-run tabletop exercise without the need for a custom game? It's a fair challenge, but proving that value definitively would require empirical study.
- Integration Difficulty with Formal Methods: How do the outputs of a game feed into formal risk management? It might be tricky to move from the game world to actionable items. Traditional tools produce risk registers, action plans, etc. The game produces experiences and some performance metrics. It will rely on effective debriefing and translation. There's a risk that insights remain tacit or forgotten, especially if there's a leadership change or time passes. Essentially, institutionalizing the insights from a gamified approach into organizational policy or procedure may be a non-trivial step that needs careful handling.

1.6.3 Usability Compared to Traditional Tools

Traditional risk tools (like risk matrices, heat maps, fault trees, etc.) are generally analytical, static, and document-focused. Our approach is experiential, dynamic, and interaction-focused. Depending on the context, one or the other might be more usable:

- In terms of ease of understanding, a well-designed game can be more immediately understandable than a complex risk register. One can learn how to play in a few minutes, whereas understanding a thick risk report might take hours. So for engaging broad audiences, our approach is more usable. On the flip side, if someone wants precise information (which risks are red vs green), a matrix is straightforward.
- Speed of setup/use: Creating a risk matrix requires workshops, data gathering, etc., but once you have it, reviewing it is quick. Running a Tetris simulation scenario might take 5–10 min per run, plus time to discuss. For a quick status update, nobody will run a simulation; they'd use a dashboard. So in day-to-day risk monitoring, traditional tools integrated with dashboards or risk software are more directly usable. Tetris-inspired simulation is more of a periodic training or exploration tool.
- User base: Traditional tools are geared for risk professionals; our game can involve a wider range of users (which is a plus for inclusivity in risk conversations). However, it might alienate those who expect a certain formality. For example, an executive might say "I don't play games, give me the bottom line." Thus, we might see different acceptance levels across organizational culture. Ideally, both could be used in tandem: the game to develop intuition and test scenarios, and the traditional tools to record and implement the findings.
- Quality of output: Traditional approaches yield tangible outputs (reports, risk lists) that can be audited, tracked, and linked to actions. Our simulation yields less tangible outputs (experience, instinct, informal knowledge). In risk management systems certified or aligned to standards, you need those tangible outputs for compliance. So in that sense, our approach is not as "usable" for compliance or reporting needs.

In practice, a hybrid approach might be best: Use the Tetris framework as a front-end learning and exploration method, then feed the insights into the back-end traditional risk management process. For example, after a simulation, the team might update their risk register to add an item "Risk of simultaneous cyber and physical outage – mitigation plan needs improvement," which they realized because in the game that scenario wrecked them. Thus, the simulation might make the traditional risk identification more robust, and also test the effectiveness of existing controls in a dynamic way.

1.6.4 Integration with Current Risk Analysis Methods

The Tetris analogy can be integrated with current methods in several ways: Scenario Planning: Traditional risk management often uses scenario analysis (qualitative narratives of future states). Our game could act as an interactive scenario. Instead of just discussing a scenario, players live through one. The results of the simulation (what went wrong or right) can then be analyzed using root cause analysis or bow-tie diagrams like any incident. In effect, we generate scenario data to feed into analysis. Over multiple runs, we could identify recurring failure modes in the game, which correspond to real failure modes worth examining via techniques like FMEA (Failure Mode and Effects Analysis).

- Stress Testing of Plans: If an organization has a contingency plan (say for a grid cyber-attack), we can attempt to map parts of that plan onto game moves. During play, we can check if players followed the spirit of the plan or diverged. Possibly the plan was not intuitive or had gaps. So the game can serve as a stress-test for existing plans or procedures. This is similar to how war games are used in military domains to test strategies.
- Augmenting Risk Registers: Risk registers list single risks often, but not combinations. We could use insights from the game to suggest which combinations or sequences of risks should be added as explicit items or at least noted. Also, if the game identifies a particular risk as consistently troublesome (like maybe one piece type always causes trouble), that might indicate that that risk is underappreciated in the formal register and needs higher priority or better controls.
- AI in Risk Monitoring: Many organizations are exploring AI for real-time risk monitoring [9] If integrated, one could feed data from actual monitoring systems into a simulation to create dynamic training that uses current real data trends as scenario seeds. Conversely, algorithms used in the simulation's AI (like anomaly detection to decide next pieces) could potentially be repurposed or inspire how to do similar detection in real operations. This is speculative, but the idea is cross-fertilization between the simulation environment and actual risk tech tools.
- Link to Standards: To satisfy frameworks like ISO 31000/31050, we can explicitly align components: For instance, ISO 31050 talks about external scanning and internal analysis [4]; in our integration we can mention that playing the game is part of an internal analysis exercise, exploring how changes (pieces) could impact objectives. If needed, one could formalize a mapping: e.g., each identified failure in the game is recorded as a risk event, with causes and consequences, and then treated just like an identified risk from a workshop.
- Training and Certification: For industries that require regular drills (like nuclear power requires emergency drills, financial sector does stress tests), the game can be integrated as one form of drill. It doesn't replace full-scale exercises, but could be a more frequent, lightweight exercise. If results are logged, organizations could maintain records to show regulators or auditors: "We regularly conduct simulation-based training on emerging risks; here are the improvements observed."

1.7 Conclusion and Future Work

In this paper, we introduced "Thinking in Tetris" as an innovative approach to managing emerging risks in critical infrastructure. We began by recognizing the increasingly complex and unpredictable risk landscape that critical infrastructures face—a world of wicked problems where traditional, linear risk management methods often fall short. To address the need for more adaptive and intuitive frameworks, we proposed the Tetris analogy, mapping the timeless puzzle game's mechanics to risk management concepts. Through this analogy, we illustrated how continuous risks (pieces) threaten a system (the board), and how agile decision-making (player moves) and effective mitigations (clearing lines) can preserve system resilience.

Acknowledging that while it has clear strengths—in fostering dynamic thinking, engagement, and holistic perspective—our approach is not a panacea. We emphasized that this tool complements rather than replaces traditional methods. We believe this balanced view is important for adoption: managers are more likely to try a new approach if it is presented as enhancing their existing toolkit, ensuring that the value created in the simulation can translate to real-world improvements (for example, updating risk registers or informing investment decisions in resilience).

The Tetris analogy for risk management contributes to the field in several ways: It provides a fresh conceptual framework for understanding emerging risks, one that is easy to grasp and communicate. It introduces a practical tool for experiential learning in risk management, which can improve preparedness and adaptive capacity of teams. It aligns with and enriches current trends (like ISO 31050's guidance on emerging risk and the push for systems thinking in resilience [3]) by offering a means to practice what those documents preach. It opens up a line of interdisciplinary collaboration—between risk analysts, cognitive scientists, and game designers—which is somewhat novel in this domain.

For future work, several avenues are promising: (1) Prototype Development and User Testing: The next step is to develop a functional prototype of the Tetris-risk simulation based on our methodology. We plan to conduct user studies, perhaps with students in critical infrastructure courses or risk management professionals, to gather feedback on the usability and learning outcomes. Measuring knowledge retention or decision-making improvement after training with the simulation would provide empirical evidence of its effectiveness [1]. Refinement of the Model: Based on testing, we may refine the mapping or rules. For example, we might introduce new game mechanics to simulate specific risk phenomena (like a "pause piece" that represents waiting for management approval, which could freeze the game momentarily with pros and cons). (2) Multi-Player Extension: As mentioned, one extension is to make it multi-player or multi-board to simulate interactions between different organizations or departments. This could better capture the negotiation and communication aspect of wicked problems. (3) Integration with Real Data: For instance, could the sequence of pieces be generated from patterns in real incident databases? If so, training could be tailored to emerging trends (e.g., more cyber pieces if cyber incidents are on the rise). Additionally, capturing how players respond could even inform AI development for automated risk mitigation systems (learning from human strategies). (4) Comparative Studies with Traditional Training: It would be useful to compare the Tetris simulation training with traditional tabletop exercises or seminars. Does the game approach lead to better retention or faster reaction times in real exercises? Such studies would help quantify the value-add of gamification in this context, contributing to research on serious games for crisis management. (5) Industry-Specific Customization: Different sectors might customize the analogy. For example, a financial risk Tetris might have pieces represent market shocks and the board represent a portfolio's risk capacity. Future work could explore variants of the game for various domains (finance, healthcare, etc.) and whether the core principles hold or need adjustment.

In conclusion, the Tetris-inspired framework appears to be a valuable complement to existing risk management practices. It excels in areas where traditional methods are weak: dealing with dynamics, interactions, and human factors. It does not replace the need for rigorous analysis and documentation, but it can feed into those. The success of integration will depend on organizational open-mindedness and careful facilitation. It requires risk managers to embrace a bit of play and uncertainty themselves, which is a cultural shift. But as newer generations enter management (many of whom grew up with video games), such approaches might gain more acceptance, especially if early adopters demonstrate tangible benefits.

We believe that managing emerging risks in critical infrastructure requires not only new tools but also new ways of thinking. The Tetris analogy invites risk professionals to think in terms of patterns, pace, and space—to see risk management as a dynamic, sometimes game-like process where the objective is to keep the system robust amidst an onslaught of challenges. By training in this mindset, critical infrastructure managers and teams can develop greater agility and insight, complementing the structured methods they already use. The positive reception of similar gamified approaches in disaster risk reduction and crisis simulation gives us confidence that this approach can gain traction [9].

References

1. ISRM Global Journal 2024—wicked problems—ISRM. https://www.theisrm.org/isrm-global-journal-2024-wicked-problems/
2. Advancing the complex adaptive systems approach to enterprise risk management with quantified risk networks (QRNs)| Scientific reports. https://www.nature.com/articles/s41598-024-71764-x
3. Infrastructure as a wicked complex process—Arizona State University. https://asu.elsevierpure.com/en/publications/infrastructure-as-a-wicked-complex-process
4. ISO 31050: Introducing risk intelligence cycle for emerging risks—WTW. https://www.wtwco.com/en-us/insights/2024/02/iso-31050-introducing-the-risk-intelligence-cycle-for-emerging-risks
5. theirm.org, https://www.theirm.org/media/9230/charities-sig-an-introduction-to-emerging-risks-and-how-to-identify-them.pdf

6. ISO 31050 Guidance for managing emerging risks to enhance resilience. http://committee.iso.org/sites/tc262/home/projects/ongoing/iso-31022-guidelines-for-impl-2.html
7. thecroforum.org, https://www.thecroforum.org/wp-content/uploads/2023/11/Breaking-Point_Critical-Infrastructure-Disrupted.pdf
8. Solinska-Nowak A, Magnuszewski P, Curl M, French A, Keating A, Mochizuki J, Liu W, Mechler R, Kulakowska M, Jarzabek L (2018) An overview of serious games for disaster risk management–Prospects and limitations for informing actions to arrest increasing risk. Int J Disaster Risk Reduct 31:1013–1029
9. Serious game launches to advance emergency responses to public health threats. https://www.nyas.org/press-release/serious-game-launches-to-advance-emergency-responses-to-public-health-threats/

Chapter 2
Facilitating the Shift to Sustainable, Secure, Clean and Affordable Energy Solutions: Regional Challenges and Geopolitical Drivers of Decarbonization in the Western Balkans

Ivana Vojinović and Andrija Leovac

Abstract The global energy transition is no longer solely a matter of technological advancement or market reorientation—it is a structural transformation shaped by climate imperatives, geopolitical volatility and institutional asymmetries. This paper examines the drivers, mechanisms and barriers to decarbonization in the Western Balkans, situating regional dynamics within broader global and European frameworks. It draws on the evolving role of the European Union's climate governance instruments, including the European Green Deal, Fit for 55, REPowerEU and the Carbon Border Adjustment Mechanism. At the same time, it explores the implications of the mentioned documents to the fossil-dependent economies. Special attention is given to the political economy of transition finance, the institutional preconditions for regulatory convergence and the asymmetric exposure of the Western Balkans to external decarbonization pressures. The analysis underscores how energy transformation in the region depends not only on access to capital and technology, but on sustained investment in governance, credibility, and social legitimacy. The paper concludes with policy recommendations that emphasize the need for differentiated integration pathways, regional coordination and institutional development as foundational elements of a just and resilient transition.

Keywords Energy transition · Decarbonization · Western Balkans · Energy security · Renewables · EU climate policy

---

I. Vojinović (✉)
Faculty of Politechnics, University of Donja Gorica, Podgorica, Montenegro
e-mail: ivana.vojinovic@udg.edu.me

A. Leovac
Faculty of International Economics, Finance and Business, University of Donja Gorica, Podgorica, Montenegro
e-mail: andrija.leovac@udg.edu.me

S. Kentera et al. (eds.), *Building Resilience in a Dynamic Critical Infrastructure Landscape*, NATO Science for Peace and Security Series C: Environmental Security,
https://doi.org/10.1007/978-3-032-04383-2_2

2.1 Introduction

The global energy system is undergoing a historic transformation shaped by the accelerating climate crisis, geopolitical instability and technological disruption. While energy has traditionally been approached as a commodity governed by market dynamics, the twenty-first century imposes a broader interpretation where sustainability, security, affordability, and system resilience are increasingly interconnected.

Achieving the goals of the Paris Agreement and the United Nations' Sustainable Development Goals (SDGs), particularly SDG 7 on clean and affordable energy, requires a fundamental restructuring of the energy economy. This entails a transition away from fossil fuels toward low-emission sources, alongside massive investment in energy efficiency, infrastructure modernization and grid stability. The International Energy Agency (IEA) projects that global demand for coal, oil, and natural gas will fall by 90%, 75%, and 55%, respectively, by 2050 under its Net Zero Emissions scenario [1]. However, the transition must be implemented in a manner that ensures reliable access to energy, mitigates socio-economic inequalities and builds resilience to external shocks. The Russian invasion of Ukraine, as well as escalating tensions in other regions, have highlighted the strategic vulnerability of the energy systems and reaffirmed energy security as a national, regional and global imperative.

This paper explores the key drivers and barriers of the clean energy transition in the Western Balkans, situating them within the broader global and European decarbonization agenda. It draws on recent policy frameworks such as the European Green Deal (EGD), REPowerEU and the Carbon Border Adjustment Mechanism (CBAM), while also considering the structural and political constraints faced by small, fossil-dependent economies. Special attention is given to the socio-political aspects of a just transition, the need for institutional reform and the role of regional cooperation in ensuring policy coherence and energy resilience.

2.2 Global Energy Transition and Decarbonization Trends

The structure of the global energy system is increasingly incompatible with the dual imperatives of climate stability and geopolitical resilience. Despite intensifying efforts to shift toward low-emission sources, fossil fuels continue to dominate both the primary energy mix and embedded infrastructure across the globe. Current daily energy consumption amounts to approximately 1.6 terajoules (TJ), and projections suggest a further rise to 1.8 TJ by 2050, even in pathways that assume robust efficiency gains [1, 2].

While the growth of renewables is accelerating (over 560 GW of new capacity was added globally in 2023), largely driven by solar photovoltaics, the system remains fossil-dependent. Coal remains to be a central pillar in electricity generation for numerous economies, with more than 40% of electricity worldwide still generated from this source [1]. The persistence of fossil-based energy systems is not only a function of technological inertia but also of economic structures that continue to externalize environmental and social costs [3].

According to the IEA, a net-zero trajectory requires an unprecedented decline in fossil fuel demand over the next two decades. This includes reductions of 90% in coal, 75% in oil, and 55% in natural gas use, along with large-scale deployment of renewable energy, storage systems and electrified end-use technologies [1]. Achieving this transition is not primarily a question of technical feasibility but one of political economy—how entrenched interests are contested, how risk is distributed, and how institutions adapt to manage disruption.

2.2.1 *Electricity Security and Infrastructure Stress*

The decarbonization of the energy sector is intrinsically tied to the electrification of economic systems. As electricity becomes the dominant vector for energy use—powering mobility, thermal processes and industry, its stability and accessibility take on new strategic significance. However, the rising share of variable renewable energy in power systems introduces temporal volatility that most existing grids are ill-equipped to manage [1, 4].

The growing mismatch between production and demand peaks exposes structural deficiencies. Many national systems lack both the balancing capacity and real-time digital infrastructure required to absorb high levels of intermittent supply. Seasonal storage remains technologically constrained, while investments in grid expansion lag behind policy ambition. IEA estimates suggest that global investment in grid systems must more than double by 2030 to avoid system-level instability [4]. Recent disruptions, including climate-related blackouts, cyber incidents, and cross-border trade interruptions, have exposed how energy systems are now entangled with national security and economic sovereignty. As such, electricity is no longer simply a technical subsystem of modern economies, but a precondition for economic continuity, public trust and strategic autonomy.

2.3 European Union's Climate Governance

The European Union (EU) has placed decarbonization at the center of its political economy, integrating climate targets with energy market reform and security strategy. The European Green Deal (EGD), supported by the Climate Law (Regulation 2021/1119) and implemented through the Fit for 55 package, commits the EU to reducing net greenhouse gas emissions by at least 55% by 2030 and achieving climate neutrality by 2050 [5].

This governance model has evolved into a strategic doctrine. Climate and energy policy now serve a dual function: they reduce emissions and enhance the Union's resilience in an unstable geopolitical context. The adoption of REPowerEU in 2022 marked a major acceleration of this shift [6]. The plan outlines measures to reduce dependency on Russian fossil fuels and to restructure the EU energy system through renewable deployment, energy savings and diversification of imports [6]. By the end of 2023, Russian gas imports were reduced by over 80%, as LNG capacity, storage coordination and interconnection mechanisms expanded across the Union [1]. REPowerEU allocates over €300 billion for these efforts, channeling funds into cross-border infrastructure, hydrogen production and grid modernization [6]. According to IEA's World Energy Investment 2024, global electricity grid investment reached approximately $400 billion in 2024, reflecting this alignment between industrial policy and clean energy targets [1].

Internal policies are increasingly supported by market instruments. The EU Emissions Trading System (ETS) has become the backbone of the Union's carbon pricing model. Its expansion to new sectors and the gradual removal of free allowances reinforce the cost signal for industrial emitters [5]. However, internal consistency alone is insufficient. The EU has begun to export its climate governance model beyond its borders through mechanisms such as CBAM.

2.3.1 The Role of CBAM in Externalizing EU Standards

The Carbon Border Adjustment Mechanism links EU climate objectives with external trade. From 2026, importers of electricity, steel, aluminum, fertilizers, hydrogen and cement will be required to pay a carbon cost equivalent to the ETS benchmark [7]. This mechanism is designed to prevent carbon leakage while promoting policy alignment among EU trading partners [7].

CBAM operates by formalizing asymmetries in emissions performance into economic obligations. The mechanism does not replicate tariffs but redefines market access based on environmental criteria. In doing so, it externalizes the EU's regulatory standards and projects internal climate logic into its trade architecture [7, 8]. For non-EU countries, including those in the Western Balkans, CBAM introduces a compliance structure that intersects with industrial policy and accession logic. In these economies, characterized by coal-heavy energy systems and limited carbon accounting infrastructure, CBAM may also incentivize reform. If accompanied by transitional support, it can trigger investments in Monitoring, Reporting and Verification (MRV) systems, National Energy and Climate Plans (NECP) implementation and domestic carbon pricing [2, 6, 9]. However, the absence of the full transposition of the EU's Third and Fourth Energy Packages into national systems, due to the policy derogations, risks creating a double periphery: countries are excluded from the benefits of full EU integration while being exposed to its regulatory costs through mechanisms like CBAM.

2.3.2 Governance Instruments and Conditional Integration

Beyond regulatory mechanisms like CBAM, the EU employs a broader suite of instruments to align the energy and climate governance of its neighbors with internal objectives. Among the most critical is the system of National Energy and Climate Plans, which serve as the formal interface between Member States, accession candidates and the EU's long-term climate strategy [6]. Although originally designed for internal coordination, NECPs have become increasingly relevant for the Western Balkans, which are encouraged to submit voluntary plans aligned with the EU framework.

The Instrument for Pre-Accession Assistance (IPA) III and the Economic and Investment Plan for the Western Balkans (adopted in parallel with the Green Agenda for the Western Balkans (GAWB)) are now partially structured around GAWB priorities, meaning that energy reform is becoming embedded in conditionality logic [6]. This reflects a deeper evolution in how the EU manages integration: not only through normative convergence, but through performance-driven access to financial and regulatory instruments.

The challenge, then, is not only to adopt EU rules but to embed them in functioning governance ecosystems. Without this, formal compliance may mask substantive divergence.

2.4 Core Drivers of Energy System Restructuring

The transformation of energy systems is increasingly governed by the interaction between market instruments, technological pathways and institutional capabilities. While solar and wind technologies dominate the expansion of installed capacity, the success of the clean energy transition depends on the capacity of states and regions to restructure governance, correct market distortions and coordinate long-term investment cycles.

2.4.1 Carbon Pricing and Strategic Market Signals

Price formation is a foundational mechanism in realigning energy systems. Yet, more than half of global CO_2 emissions remain unpriced or insufficiently priced to support decarbonization trajectories [1]. The European Union's Emissions Trading System represents the most advanced cap-and-trade scheme globally, but its expansion to new sectors and gradual reduction of free allowances has exposed tensions between competitiveness and climate ambition [5].

Beyond the EU, carbon pricing remains fragmented and politically contested. IEA projections suggest that aligning carbon prices with net-zero goals could reduce global emissions growth by 50% in the power and industrial sectors by 2035 [1].

However, carbon pricing alone cannot compensate for structural inertia. It must be embedded in a broader policy matrix that includes regulatory standards, green procurement and targeted subsidies. These insights are reinforced by global decarbonization studies, which stress that effective price signals require long-term policy credibility and integration with broader fiscal systems [10]. Fragmented pricing schemes without clear future trajectories fail to generate the financial certainty needed for large-scale investment in carbon–neutral technologies.

Price signals are most effective when paired with credible transition frameworks. The presence of predictable carbon pricing enables investors to allocate capital away from fossil fuel assets and toward long-duration storage, hydrogen and electrified industrial processes. Without these signals, financial risk is misallocated and low-carbon innovation is delayed.

2.4.2 Technology Diversification and Infrastructure Adaptation

The transition to net-zero is constrained by the current technological portfolio of most economies. While renewables now account for the majority of new power capacity additions, the decarbonization of industry, heating and heavy transport remains highly dependent on emerging solutions.

Green hydrogen is a key vector in decarbonizing hard-to-abate sectors, including steelmaking and ammonia production. However, its cost remains prohibitive and supportive infrastructure is limited to pilot-scale deployment. The European Union's Hydrogen Accelerator and Hydrogen Backbone initiatives aim to build continental-scale electrolyzer capacity and repurpose existing gas pipelines for cross-border hydrogen flows [6]. Carbon capture and storage (CCS) has re-entered mainstream policy discussions, particularly for emissions from cement, chemicals and refineries. Yet, deployment is concentrated in a few jurisdictions and often lacks the regulatory clarity and liability frameworks needed to attract investment. Moreover, concerns remain over its cost-effectiveness and long-term environmental risks [10]. Grid infrastructure also represents a major constraint. According to IEA, global investment in transmission and distribution networks must double by 2030 to accommodate variable renewables and rising electrification levels [4]. Delays in permitting, outdated planning procedures, and a lack of regional coordination have slowed progress.

In emerging economies, structural constraints such as limited fiscal space, high risk premiums and governance bottlenecks further restrict deployment of advanced energy technologies [10]. These are not merely technical problems, on contrary they reflect deeper institutional path dependencies that shape how policy is implemented and how public trust is cultivated. The challenge is systemic. Technologies must be deployed in parallel with institutional reform. Without adaptive planning, digital monitoring and cross-border coordination, even the most advanced technologies will deliver suboptimal outcomes. The core drivers of the energy transition, therefore, are not only technological and financial, they are deeply institutional.

2.5 Regional Challenges in the Western Balkans

The Western Balkans confront a layered and persistent set of challenges in aligning with global decarbonization frameworks. Although coal intensity, energy poverty and infrastructure obsolescence are well-known issues, the deeper constraints are structural: institutional fragmentation, weak regulatory capacity and limited fiscal space for transition investment [11].

Most countries in the region continue to depend on lignite-based power, justified politically through arguments of energy sovereignty, but economically sustained through suppressed pricing, regulated markets and the absence of carbon cost internalization [1, 12, 13]. According to the Energy Community Secretariat, no Contracting Party had implemented an operational carbon pricing mechanism by the end of 2023 [13]. This institutional inertia is compounded by inefficiencies in energy use. The region's final energy consumption per unit of GDP is significantly above EU levels, while distribution losses can reach up to 20% in some networks [14]. Grid expansion and renewable integration projects face chronic underinvestment, due to a combination of high sovereign risk, administrative delays and constrained local co-financing capacities [15]. Although formal structures like the NECPs exist in many cases, implementation is weak. Submissions often lack robust modeling, funding strategies or binding measures [6, 9, 13]. These institutional limitations echo broader findings from global decarbonization literature: that successful transition depends not only on having plans, but on embedding them in capable and coherent public systems [16]. Carbon pricing, which plays a central role in EU climate governance, remains politically marginalized in the Western Balkans. Without functioning MRV systems and with limited public trust, such instruments are often perceived as externally imposed. The World Bank and the International Monetary Fund have emphasized that successful carbon pricing in emerging markets requires alignment with macroeconomic policy, redistributive mechanisms and transitional compensation schemes [12, 15]. Also, CBAM introduces a new vector of pressure. Even in its transitional phase, CBAM forces local producers to internalize carbon metrics, often for the first time, despite having no domestic carbon market architecture [7, 15]. This asymmetry deepens the divergence between internal EU decarbonization logic and external partners' institutional readiness [11, 16].

Most critically, transition in the Western Balkans is undercut by weak fiscal capacity. Transition-aligned investments require long-term public borrowing frameworks, public–private partnerships, and green finance institutions—none of which are systematically embedded in the region [15]. Without these, the transition risks becoming a donor-driven rather than a sovereign developmental strategy. Global frameworks stress that these gaps are not peripheral, they are central. According to the IPCC's Special Report on Global Warming of 1.5 °C, institutional capacity is one of the most crucial enabling conditions for decarbonization pathways [11]. In its absence, even well-designed policies fail to produce systemic transformation.

2.6 Conclusion

The Western Balkans' decarbonization pathway is situated within a highly asymmetric geopolitical and economic landscape. While global climate imperatives and EU integration frameworks impose increasing pressure for structural reform, the region's institutional and fiscal constraints create a persistent gap between ambition and implementation. The energy transition here is not merely a question of infrastructure deployment or emissions reduction. It is a test of institutional resilience, political commitment and social cohesion.

The EU's policy architecture, through instruments such as Fit for 55, REPowerEU, and CBAM, has accelerated the externalization of its climate governance. Yet, these mechanisms presume a level of institutional maturity and fiscal capacity that the Western Balkans does not currently possess.

Bridging this structural asymmetry requires a recalibration of the integration logic. The success of the region's transition hinges not only on regulatory approximation but on sustained investment in administrative capacity, long-term planning frameworks and multi-level governance coordination [13, 14]. Technical assistance must evolve from fragmented project delivery to embedded institutional partnerships. Capacity-building should prioritize the professionalization of public administration, digitalization of energy data systems and enhancement of cross-sectoral coordination. Fiscal architecture must also be realigned. Transition finance needs to be decoupled from ad hoc grant dependency and embedded in national budget frameworks. This entails expanding fiscal space through green bond issuance, concessional lending facilities and region-specific climate investment platforms that reduce perceived risk for both domestic and international investors [13]. Coordinated donor action should be aligned with national NECPs, while policy-based lending could serve as a lever for governance reform.

Finally, the social dimension of the transition must not be secondary. Energy affordability, employment restructuring, and territorial cohesion are key to maintaining political legitimacy. Just transition mechanisms must be operationalized not only through EU instruments such as the Social Climate Fund, but through domestic policies that embed public participation, retraining schemes and local development planning [11]. In sum, the Western Balkans require a decarbonization strategy that is both technically ambitious and institutionally realistic. The alignment with EU climate architecture must be matched by a recognition of local constraints and by a coordinated effort to transform them.

References

1. International energy agency (IEA): World energy outlook 2024. IEA, Paris (2024)
2. Vojinović I (2023) Cost-benefit analysis of adopting and implementing the EU environmental acquis in Montenegro. Doctoral Thesis, University of Donja Gorica, Podgorica
3. World economic forum (WEF): Fostering effective energy transition 2023. WEF, Geneva (2023)
4. International energy agency (IEA): Electricity 2024: analysis and forecast to 2026. IEA, Paris (2024)
5. European commission: Fit for 55—Delivering the EU green deal. Publications office of the European Union, Brussels (2021)
6. European commission: REPowerEU Plan. Publications office of the European Union, Brussels (2022)
7. European parliament and council: Regulation (EU) 2023/956 Establishing a carbon border adjustment mechanism. Off J Eur Union L130, pp 52–87 (2023)
8. Siddi M (2020) The European green deal: assessing its current state and future implementation. In: Finnish institute of international affairs working paper 113. FIIA, Helsinki
9. Goldthau A (2017) Rethinking the governance of energy infrastructure: scale, ideology and politics. Energy Res Soc Sci 1:134–140
10. Vojinović I (2024) Understanding main drivers of global decarbonization. In: Badnjević A, Gurbeta Pokvić L (eds.) MEDICON'23 and CMBEBIH'23. IFMBE proceedings, vol 93, pp 748–756. Springer, Cham
11. Rogelj J et al (2018) Mitigation pathways compatible with 1.5 °C in the context of sustainable development. In: IPCC special report on global warming of 1.5 °C, Cambridge University Press, Cambridge
12. International monetary fund (IMF): A proposal for an international carbon price floor among large emitters. IMF Working Paper WP/21/254, Washington, DC (2021)
13. Energy community secretariat: Implementation report 2023. Energy Community, Vienna (2023)
14. Energy community secretariat: Monitoring report on electricity market integration in the Western Balkans (2023)
15. World Bank: Financing decarbonization in emerging markets. World Bank Group, Washington, DC (2022)
16. OECD: Institutional capacity and the climate transition: challenges in governance and coordination. OECD Publishing, Paris (2021)

Chapter 3
Critical Underwater Infrastructures: New Challenges and Threats

Alessandro Lazari

Abstract Critical Underwater Infrastructures (CUIs)—including submarine cables and pipelines—constitute indispensable components of global connectivity, energy transport, and digital resilience. Despite their foundational role in international communication and commerce, these infrastructures remain highly vulnerable to a broad spectrum of risks, including environmental hazards, systemic degradation and deliberate sabotage. Recent high-profile incidents, such as the Nord Stream pipeline explosions and disruptions to submarine cable systems in Northern Europe, have revealed the fragile nature of these assets and the disrupting consequences of their malfunction or destruction. This paper offers a multidisciplinary assessment of CUIs, examining their technical features, risk exposure and role within global security architectures. Special attention is given to emerging hybrid threats, cross-domain interdependencies, and the capabilities of monitoring and deterrence technologies—including AI-enhanced surveillance and unmanned systems. The analysis integrates perspectives from existing legal instruments (e.g., UNCLOS) and proposes actionable pathways to enhance detection, resilience, and cooperative protection of CUIs in contested and high-risk maritime environments.

Keywords Critical underwater infrastructures · Submarine cables · Offshore pipelines · Hybrid threats · Maritime security · Infrastructure resilience · NATO · UNCLOS · Cyber-physical systems

A. Lazari (✉)
Department of Engineering for Innovation, University of Salento, Lecce, Italy
e-mail: alessandro.lazari@unisalento.it

S. Kentera et al. (eds.), *Building Resilience in a Dynamic Critical Infrastructure Landscape*, NATO Science for Peace and Security Series C: Environmental Security,
https://doi.org/10.1007/978-3-032-04383-2_3

3.1 Introduction

The security and resilience of infrastructure systems have become central themes in contemporary geopolitical, technological, and strategic discourses. Among these, *Critical Underwater Infrastructures* (CUIs)—notably submarine telecommunications cables and energy pipelines—constitute a largely invisible yet indispensable backbone of the global economy. These infrastructures facilitate 95% of international data traffic, support the export and import of hydrocarbons, and form the digital and energetic layer of global interdependence [1, 2]. Despite this criticality, CUIs remain insufficiently addressed in both national defence postures, leaving them exposed to an evolving landscape of complex threats.

Recent geopolitical crises and unexplained underwater incidents have brought this vulnerability under brighter focus. The 2022 sabotage of the Nord Stream pipelines in the Baltic Sea and the recurrent disruptions to transcontinental submarine cable systems exemplify a pattern of strategic targeting with hybrid characteristics—blurring conventional boundaries between physical sabotage, cyber intrusion, and information operations [13, 17]. These cases illustrate how CUIs are becoming both high-value targets and potential triggers for escalation [3].

From a security policy perspective, CUIs present a unique challenge due to their cross-jurisdictional nature and their integration into both civilian and military systems. Traditional maritime security strategies prove insufficient when faced with transboundary cables, undersea assets and actors capable of acting below the threshold of armed conflict [9].

This paper adopts a global and multidisciplinary lens to examine CUIs across four dimensions: (i) the evolving threat landscape, including the emergence of hybrid and cyber-physical risks; (ii) the enabling technologies for monitoring and resilience, particularly the role of AI and unmanned systems; (iii) the applicability and limitations of existing legal and policy frameworks such as UNCLOS, NATO resilience doctrine, and regional maritime strategies; and (iv) strategic recommendations aimed at enhancing the detectability, resilience, and coordinated protection of CUIs in both peacetime and crisis scenarios.

By reframing CUIs not only as infrastructure assets but as security-relevant systems embedded in a multipolar and contested maritime domain, this study contributes to a more integrated approach to maritime security, critical infrastructure protection, and hybrid threat deterrence.

3.2 Evolving Threat Landscape for Critical Underwater Infrastructures

The threat environment surrounding Critical Underwater Infrastructures (CUIs) has grown increasingly complex. These infrastructures are exposed to multidimensional hazards that span from environmental degradation to state-sponsored sabotage. This section provides a structured taxonomy of risks facing CUIs, emphasizing their cumulative and often interdependent nature [1].

3.2.1 Natural Hazards

CUIs are inherently exposed to risks due to their deployment on or beneath the seabed, often in seismically or climatically active regions. Submarine cables and pipelines are vulnerable to underwater earthquakes, seabed shifts, volcanic activity and submarine landslides. These phenomena can damage cables or pipelines, causing extensive disruptions and ecological damage. Cyclones and storms, intensified by climate change, have the potential to damage shallow-water CUIs through coastal erosion, wave action and sediment displacement [3, 6]. The long-term effects of sea-level rise further improve stress on older infrastructure systems, particularly those in littoral zones.

3.2.2 Accidental Damage

Human activities in maritime domains—whether legal or illicit—pose frequent and unpredictable risks to CUIs. Fishing trawlers, maritime construction, seabed mining and anchoring by commercial vessels are leading sources of unintentional damage [7]. Incidents involving poorly mapped zones have demonstrated how insufficient situational awareness can result in service disruptions. In heavily trafficked corridors, the risk of mechanical entanglement or collision with submerged assets remains elevated.

3.2.3 Intentional Disruption and Hybrid Threats

The strategic relevance of CUIs has increasingly made them targets of deliberate interference by both state and non-state actors. Hybrid threats have introduced new modalities of attack. These include the covert placement of explosives, cyber-physical manipulation of remotely operated systems and electronic interference and communications sniffing at cable landing stations or nodes.

The Nord Stream pipeline sabotage in 2022 marked a paradigmatic shift in the understanding surrounding CUIs, showcasing how relatively deniable actions below the threshold of conventional war can lead to disproportionate geopolitical consequences [11, 15].

Moreover, information-poisoning and propaganda operations can induce economic panic or undermine confidence in national or regional resilience postures.

In such contexts, the exploiting of both physical and cognitive effects highlights the multi-layered dimension of hybrid threat.

3.2.4 *Geopolitical and Legal Tensions*

CUIs often cross Exclusive Economic Zones (EEZs), disputed waters, or international seabed areas. As a result, their protection suffers from fragmented jurisdiction, limited enforceability and gaps in accountability. Tensions over maritime boundaries—such as those in the Eastern Mediterranean, South China Sea or Arctic—frequently involve or imply the existance or future development of strategic underwater assets.

3.2.5 *Infrastructure Aging and Maintenance Deficits*

A large proportion of global submarine cables and offshore pipelines were laid more than two decades ago and are approaching their operational lifespans. These aging systems suffer from material fatigue, corrosion and outdated monitoring systems. The failure of CUIs can have cascading effects on dependent sectors and the effects could be particularly severe in small states or remote regions reliant on single points of connectivity or energy supply.

3.3 Technological Enablers for the Protection of Critical Underwater Infrastructures

The protection of Critical Underwater Infrastructures (CUIs) increasingly relies on the deployment of advanced technologies that enable persistent monitoring, anomaly detection and rapid response across vast and often inaccessible maritime domains. Given the complexity of the operational environment, a multi-layered technological architecture has emerged as core of infrastructure resilience. This section outlines the principal technological capabilities currently available or under development for the security of CUIs.

3.3.1 Satellite-Based Surveillance and Maritime Domain Awareness

Satellite remote sensing plays a foundational role in detecting anomalous activities in the vicinity of CUIs, particularly in areas lacking persistent naval presence. Synthetic Aperture Radar (SAR) systems, capable of generating high-resolution images independent of cloud cover or daylight, are critical for identifying vessel loitering, unauthorized seabed operations and signs of physical disturbances near infrastructure routes [5, 12].

New generation of sensors contribute further by detecting environmental changes—such as turbidity or leaks—that may signal damage or sabotage.

Moreover, Automatic Identification System (AIS) receivers allow for real-time tracking of maritime traffic, supporting the identification of vessels operating in proximity to sensitive infrastructure, including those transmitting spoofed or absent identifiers.

These developments now offer the potential for near-continuous coverage of strategic maritime corridors, enabling not only post-event forensics but also anticipatory anomaly detection.

3.3.2 Unmanned Underwater Vehicles (UUVs) and Remotely Operated Vehicles (ROVs)

Autonomous underwater systems constitute the most direct inspection and intervention tools available for CUIs. Equipped with sonar, optical, and magnetic sensors, Unmanned Underwater Vehicles (UUVs) and Remotely Operated Vehicles (ROVs) are deployed for infrastructure mapping, damage assessment, and localized surveillance [4, 5].

In military and strategic contexts, armed or sensor-equipped UUVs may also be tasked with deterrence missions—patrolling critical nodes or acting as early-warning sentinel. On the long term, the integration of AI algorithms for path optimization and underwater recognition will enhances the autonomy and effectiveness of such systems.

3.3.3 Distributed Sensor Networks and Acoustic Monitoring

A promising evolution in CUI security involves the deployment of distributed sensor along critical routes. Fiber-optic cables can be repurposed for Distributed Acoustic Sensing (DAS), detecting vibrations or pressure anomalies resulting from nearby movement, drilling or explosive placement [10, 12].

Hydrophones and passive acoustic sensors complement these systems by detecting acoustic signatures of submersibles or diver.

Finally, advanced signal processing enables the classification of threats based on acoustic fingerprints, contributing to early warning and forensics.

Integration with shore-based fusion centers ensures that such alerts are contextualized and enriched with satellite, AIS, and intelligence data, enabling decision-makers to act within operationally relevant and effective timeframes.

3.3.4 Cyber-Physical Security Integration

CUIs are increasingly managed through Supervisory Control and Data Acquisition (SCADA) systems and other industrial control networks, making them susceptible to cyber-physical attacks [8, 17]. Intrusion into control interfaces can disable alarms, interfere on status readings, or remotely manipulate flow control in pipelines, potentially coordinating or hiding physical attacks.

As result, CUI protection must encompass robust cybersecurity measures, including network segmentation, intrusion detection, encryption, and endpoint protection for control systems. The convergence of cyber and physical threat vectors also requires a shift in incident response protocolsv—linking maritime domain awareness with cybersecurity situational awareness.

3.3.5 Artificial Intelligence and Predictive Analytics

AI-enabled systems are increasingly being adopted to fuse large volumes of heterogeneous data into coherent risk assessments. Machine learning models can be trained to detect patterns of pre-attack behaviour, such as the anomalous presences near cable landings or subtle indicators of active reconnaissance by potential antagonists [5, 14].

Predictive maintenance algorithms and digital twins, applied to pipeline stress data and historical failure records, can forecast degradation timeframes, thus allowing for preventive interventions, before physical damage manifests.

These technological advancements, while powerful enablers, are not a panacea. Their effectiveness depends on the presence of interoperable architectures and strategic coordination between public, private and often military actors. In the absence of these factors, even the most advanced technologies risk becoming poorly integrated and leading to the expected results in terms of security and resilience.

3.4 Legal and Institutional Frameworks for the Protection of Critical Underwater Infrastructures

The protection of Critical Underwater Infrastructures (CUIs) is shaped by a combination of legal instruments and institutional mechanisms spanning international law, regional security arrangements, and national regulatory regimes [2]. This section reviews the primary frameworks relevant to CUI protection.

3.4.1 United Nations Convention on the Law of the Sea (UNCLOS)

UNCLOS remains the cornerstone of international maritime law, codifying the rights and responsibilities of states in relation to territorial waters, exclusive economic zones (EEZs) and the high seas [16].

However, UNCLOS provides only limited enforcement mechanisms and does not establish binding obligations for the active protection of CUIs. While it criminalizes intentional damage to submarine cables, it does not define or address hybrid threats, sabotage below the threshold of armed conflict or the protection of digital infrastructure carried by such cables. Furthermore, its reliance on state-level implementation creates inconsistencies in enforcement capacity across jurisdictions.

In disputed maritime areas or regions lacking stable governance, the legal protections afforded by UNCLOS remain theoretical. This asymmetry is particularly concerning in regions where strategic CUIs intersect with zones of contested sovereignty or competition.

3.4.2 NATO Policy and Strategic Concept

NATO has increasingly recognized the strategic relevance of CUIs as part of its broader focus on infrastructure resilience, hybrid threat deterrence, and the protection of critical domains. The 2022 NATO Strategic Concept explicitly identifies the need to strengthen resilience against threats to critical infrastructure, including those that underpin cyber and energy domains [11, 13].

Key instruments supporting this agenda include:

- Counter-Hybrid Support Teams (CHSTs), which offer tailored assistance to Allies facing complex, blended threats.
- Exercises such as Dynamic Mongoose and Steadfast Defender, which test undersea warfare and protection protocols.

- NATO's Emerging and Disruptive Technologies (EDT) strategy, which encourages the adoption of AI, autonomous systems, and space technologies in infrastructure protection.

Nonetheless, NATO's capacity to intervene directly in CUI protection is limited by the principle of national sovereignty, therefore the development of interoperable surveillance systems, joint maritime domain awareness frameworks and resilience benchmarks remains an area for further institutional development.

3.4.3 European Union Directives and Programs

In recent times, after almost two decades of work exclusively in the domain of protection of critical infrastructures, the European Union has adopted a broader approach to the security and resilience of essential services and vital assets, integrating physical, cyber, and hybrid dimensions through the following instruments:

- The Directive on the Resilience of Critical Entities (CER Directive, 2022),
- The NIS2 Directive (2022), and
- The overarching European Programme for Critical Infrastructure Protection (EPCIP).

These instruments require Member States to identify critical infrastructure operators, conduct risk assessments, and implement incident response and continuity plans. Importantly, they mandate a cross-border perspective and encourage cooperation with third countries—an essential feature given the global nature of CUIs [7, 8].

The current legal and institutional architecture offers a foundational but fragmented approach to CUI protection. To meet the demands of the evolving threat environment, future efforts should focus on enhancing interoperability, expanding the scope of multilateral agreements to explicitly include CUIs, and harmonizing legal definitions and operational protocols across jurisdictions.

3.5 Strategic Recommendations for Enhancing the Protection and Resilience of CUIs

The increasing exposure of Critical Underwater Infrastructures (CUIs) to both traditional and hybrid threats necessitates a recalibration of strategic priorities, technological investment, and international cooperation. While recognition of their criticality has grown, practical mechanisms for their protection remain inconsistent across regions and fragmented across legal domains. Based on the preceding analysis, this section outlines a potential way for advancing the resilience and protection of CUIs at both national and international levels.

3.5.1 *Establish Minimum Resilience Baselines Through Multilateral Cooperation*

NATO and its partners should lead efforts to define and promote minimum resilience standards for CUIs, particularly in relation to redundancy, monitoring, incident response, and recovery. These standards could be aligned with existing principles but adapted to include the specificities of subsea infrastructure [7, 8].

3.5.2 *Promote Interoperable Response Frameworks for Infrastructure Incidents*

Most countries lack standardized procedures for cross-border response to underwater infrastructure incidents. NATO and other multilateral organizations should facilitate the development of joint operational protocols for:

- Subsea damage attribution and forensics;
- Multi-agency coordination during incidents (civil, naval, cyber);
- Decision-making thresholds for escalating from monitoring to intervention.

Tabletop exercises and simulations should be conducted with both NATO members and non-NATO coastal states in high-risk corridors.

3.5.3 *Foster Legal Clarity and Norm Development in International Waters*

UNCLOS remains foundational but insufficient to address hybrid threats or establish deterrence in ambiguous maritime scenarios. NATO and the EU—alongside the UN and key coastal states—should pursue diplomatic initiatives aimed at:

- Developing interpretive guidance on CUI protection under UNCLOS;
- Expanding norms on prohibited actions in international waters;
- Supporting capacity-building in under-resourced jurisdictions to ensure compliance and enforcement.

Where consensus would be difficult to reach, coalitions of states may promote voluntary codes of conduct, as occurred in other domains such as cyber.

3.6 Conclusion

Critical Underwater Infrastructures (CUIs) are among the most indispensable yet underprotected components of the global security and economic ecosystem. Their disruption has the potential to trigger cascading effects across continents and sectors.

The analysis presented in this paper has demonstrated that CUIs face a broad spectrum of risks that extend beyond traditional infrastructure concerns, also because of the intrinsic nature and operational context of these infrastructures that are place underwater in a very non-cooperative environment. Their vulnerability is no longer hypothetical; recent incidents have confirmed the intent and capability of both state and non-state actors to exploit them as strategic targets, while climate pressures and aging infrastructure further increase systemic exposure.

In response, the protection of CUIs must evolve from a sector-specific or national concern to a priority within the broader framework of collective security and infrastructure diplomacy. Legal frameworks such as UNCLOS offer a base but must be supplemented by emerging norms, voluntary protocols and operational doctrines that explicitly address hybrid threats and cross-domain convergence and integration. NATO's strategic posture, the EU's regulatory landscape and regional maritime strategies offer valuable foundations but require alignment, empowerment and implementation at tactical level.

Technological innovation provides unprecedented opportunities for predictive monitoring and resilient response. Yet technology alone is insufficient. Without interoperable coordination mechanisms and inclusive governance models many of these capabilities will remain reactive.

The path forward demands a commitment to increased multilateralism, contextualised resilience planning and improved legal frameworks.

As threats grow and attribution becomes more complex, the imperative to act has never been so urgent.

References

1. Bennett A (2019) Submarine cables and infrastructure vulnerabilities: threats from private and state actors. J Strat Infrastruct Stud 12(2):45–63
2. Bueger C, Liebetrau T (2021) Protecting hidden infrastructure: the security politics of the global submarine data cable network. Contemp Secur Policy 42(1):145–167
3. Bueger C, Liebetrau T (2023) Critical maritime infrastructure protection: what's the trouble? Marit Secur Rev 9(1):30–42
4. Bruzzone AG, Massei M, Maglione GL (2016) Simulation of manned & autonomous systems for critical infrastructure protection. In: Proceeding of the International Defense and Homeland Security Simulation Workshop
5. Bruzzone AG, Massei M (2023) Strategic engineering for underwater infrastructure resilience: a collaborative framework. Simul Resil Eng 8(3):101–115
6. Dvorak Z, Figuli L, Miske M (2024) Evolution and perspective of sea/ocean facilities critical infrastructure protection in Europe. Ocean Eng Rep 17(4):221–238
7. Directive on the resilience of critical entities (CER). Off J Eur Union, L333/164

8. Directive (EU) 2022/2555 on measures for a high common level of cybersecurity (NIS2). Off J Eur Union, L333/80
9. Gheorghe AV, Vamanu DV, Katina PF, Pulfer R (2018) Critical infrastructures, key resources, and key assets: risk and vulnerability analysis. Springer International Publishing
10. Kerigan-Kyrou D (2013) Critical energy infrastructure: operators, NATO, and facing future challenges. NATO Energy Security Centre of Excellence White Paper
11. Monaghan S, Svendsen O, Darrah M, Arnold E (2023) NATO's role in protecting critical undersea infrastructure. NATO Defense College Policy Brief, pp 23–07
12. Nowak PR, Gajewski T, Szlachta A (2024) Explosive materials management near maritime critical infrastructure: safe disposal considerations. MaritE Def Technol J 19(2):88–96
13. Ruehle M (2014) NATO and emerging security challenges. NATO Review Online, Winter 2014 Issue
14. Rus S, Ion A (2015) Protecting the national and European critical underwater infrastructures: strategic approaches and recommendations. Secur Stud Rev 10(3):203–225
15. Soldi G, Gaglione D, Raponi S, Forti N (2023) Monitoring of critical undersea infrastructures: the nord stream and other recent case studies. NATO Science and Technology Organization (STO) Technical Report No. TR-UX-2210
16. Wrathall LR (2010) The vulnerability of subsea infrastructure to underwater attack: legal shortcomings and the way forward. Int Law Secur J 5(1):11–33
17. Publications office of the European union (2023) Hybrid threats: a comprehensive resilience ecosystem. Luxembourg: JRC Technical Report JRC129019. https://doi.org/10.2760/37899

Chapter 4
Artificial Intelligence and Biomedical Science: Challenges and Risks

Elizabeta Ristanović, Nikoleta Đorđevski, Sonja Atanasievska Kujović, and Vesna Protić Đokić

Abstract The COVID-19 pandemic has shown that artificial intelligence (AI) has significantly improved our ability to monitor, predict and respond to such complex health crises as outbreaks (spread assessment, trajectory tracking, diagnostic tools development, drug and vaccine discovery assistance, big data processing, optimizing of resources management, simplifying of administrative procedures etc.). AI is also very useful tool for decision makers and effective medical management in crises. The potential use and benefits of AI in biomedicine (diagnostics, treatment, surgical procedures, patient monitoring, prognostic and predictive modeling) are practically unlimited. It is also opening a numerous of ethical and security issues, because healthcare today in the era of hybrid wars could be also considered as a security challenge.

Keywords AI · Health-care · Outbreaks · Date-protection · Biotechnology

4.1 Introduction

The COVID-19 p(l)andemic opened up numerous questions and represented an introduction to the geopolitical confrontations of the great powers, which became evident at the very beginning of the crisis, primarily between the USA and the People's

---

E. Ristanović (✉) · N. Đorđevski · S. A. Kujović · V. P. Đokić
Military Medical Academy, University of Defence, Belgrade, Serbia
e-mail: elizabeta.ristanovic@vma.mod.gov.rs

N. Đorđevski
e-mail: nikoleta.djordjevski@gmail.com

S. A. Kujović
e-mail: sonja.atanasievska.gmail.com

V. P. Đokić
e-mail: vesnapdj@gmail.com

S. Kentera et al. (eds.), *Building Resilience in a Dynamic Critical Infrastructure Landscape*, NATO Science for Peace and Security Series C: Environmental Security,
https://doi.org/10.1007/978-3-032-04383-2_4

Republic of China. These confrontations were related to the origin of the virus, the way of its spread, the geopolitics of vaccines, medical diplomacy etc.

Crisis management affected the outcome of election races in a large number of countries around the world, including the USA itself. The pandemic has also widely introduced artificial intelligence (AI) into our daily life and biomedical practice as well as the unsuspected possibilities of its application in response to health crises (spreading estimation, trajectory tracking, diagnostic tools, drugs and vaccines development, processing of large data amounts, prognostic and predictive modeling, optimizing of resource management, information placement, crisis management and communication) [1].

By the way, the question of simulations raised in October 2019, when a scenario of a future pandemic caused by a respiratory virus was played out at John Hopkins University, USA. It is interesting that the aforementioned university was in charge of programs for processing and storing medical data from different parts of the world during the pandemic, with the use of AI.

All of this has opened up numerous ethical and security dilemmas, issues of misuse of science, genetic engineering, biotechnology, nanotechnology, as well as AI itself, because biological and cyber security are part of the overall approach to looking at security systems today [2]. This is especially important considering the fact that health care as a whole, and especially infectious diseases, microorganisms, genetic data, as well as biological threats can be seen as a specific, extremely current security challenge. After all, the COVID-19 virus immediately after the pandemic put on the camouflage uniform of war. The issues of biological researches, risks as well as the control of genetic and biological data, has become the subject of confrontation between the great powers since the beginning of the Ukrainian crisis, related to the existence of biological laboratories in the post-Soviet space and non-compliance with the Biological Toxin Weapons Convention (BTWC) [3].

The National Biodefense Strategy and Implementation Plan was adopted in the USA in October, 2022, while in March 2024, in this country was adopted the Biosecure Act with the aim to restrict funding and genetic data use by foreign adversary biotech companies, primarily directed against the influence of Chinese biotech companies [4].

In the meantime, the AI development is taking place extremely fast and gaining more and more space in our daily life and the question is whether we will control it or it will control us? The other question is whether the man of the future will be a combination of the biological, physical and digital world.

The National Assembly of Serbia on June 13, 2024 opened a public debate on the Strategy for the AI Development until 2030. It is estimated that it will affect 40% of the activity and the question raised whether it is a real development chance or a risk. On the same day was adopted the Regulation (EU) 2024/1689 of the European Parliament establishing harmonized rules on AI.

4.2 Challenges of AI in Healthcare and Biomedicine

AI, the simulation of human intelligence processes using computer systems, affects today all segments of human life, but it has a particularly deep impact in the health sector, which, as a complex and dynamic field, requires efficiency, precision and constant innovation. Thanks to AI, it is possible to analyze huge amounts of data, recognize patterns and make predictions, all in a fraction of the time needed to make decisions, which is especially important from the aspect of management and leadership. AI in healthcare is already not just a futuristic concept, but a reality by which medical professionals diagnose, treat and monitor patients. Today's health challenges are truly numerous, and they primarily relate to population aging, chronic diseases, malignant diseases, epidemics and pandemics of infectious diseases. AI offers the promise of transforming healthcare delivery, making it more accessible, cost-effective, accurate and tailored to individual needs, at all levels—from early disease detection to personalized treatment plans. Healthcare systems around the world are adopting and implementing AI to simplify operations, reduce costs, and improve patient treatment and prognosis [5].

AI has entered all segments of healthcare, including early diagnostics and predictive modeling. AI algorithms can analyze X-ray, MR, CT images to detect abnormalities, including tumors and fractures (*Google Deep Mind* detecting eye diseases, 89% of all gene mutations that can potentially lead to the appearance of diseases in humans; *IBM Watson* usage for cancer diagnosis and treatment, *Derm Engine*—the early detection of skin cancer) [6]. AI algorithms that analyze medical images and laboratory results reduce human error and significantly increase the accuracy of diagnostic procedures. Based on the assessment of medical history, lifestyle information and genetic characteristics, we can use AI to predict potential health risks (e.g. for the occurrence of cardiovascular and cerebrovascular diseases and strokes, sepsis, diabetes, neurodegenerative diseases and other chronic conditions, including malignant ones, as well as personalized mental status assessments, all of which enable adequate prevention and timely interventions [7]. Thus, future AI-driven diagnostic tools can offer real-time analysis, including genomic data, lifestyle factors, and environmental conditions with extremely high accuracy. You can help in creating personalized treatment regimens tailored to each patient's unique needs, including targeted gene therapies for oncology, personalized prescriptions, tailored medical advice… [8]. AI can analyze complex biological data and target sequences, significantly reducing the time and costs of drug and vaccine development (e.g. Alzheimer's disease, COVID-19, new antibiotics, vaccines, oncology disease therapy…) and simplifying the clinical trial process by identifying suitable candidates, monitoring the application and analyzing the results of processing a large amount of data [9]. Various medical procedures, including robot-assisted operations and surgical interventions, extensivly use AI, which allows them greater precision and control, reduces the risk of complications and shortens recovery time (e.g. the *da Vinci Surgical System*, which has performed millions of minimally invasive surgeries worldwide). In October 2022, the People's Republic of China opened the world's first AI hospital, *"Agent Hospital"* in which the virtual patients are treated by AI generated doctors. The head of the hospital's research team, Liu Yang, said this innovative method allows real doctors to treat virtual patients and provides medical students with better training, without the risk of harming real patients. Also, in these conditions, it is possible to simulate and predict various medical scenarios, such as the spread, development and control of infectious diseases in the region [10].

With the help of *Alibaba Cloud's AI Technology*, researchers from China, Hong Kong and Australia has recently discovered as many as 162,000 new RNA viruses in various environments, including the atmosphere, hot and geothermal sources, and data on their evolutionary origins and resistance in the external environment, through RNA analysis sequences from databases as well as the structures of hidden proteins used in replication.

Virtual health assistants use AI to provide medical advice, medication reminders, and mental support to patients -*chat bots.* AI also enables remote monitoring of patients through portable-wearable devices and smart technology, enabling continuous care outside of traditional hospital environments (post-operative conditions, chronic diseases, etc.) Telemedicine AI powered platforms offer virtual consultations, reducing barriers to access to healthcare, especially in underserved areas (rural areas etc.) and also enable integration of symptoms with the electronic health record [9].

Healthcare AI is a driving force for positive change and reflects a potentially new era of medical excellence. Simplifying administrative tasks (scheduling, billing, and medical records managing) by automating them will allow healthcare professionals to pay more attention to patient care and thus further improve the treatmen efficiency [11].

With the automation of routine tasks and capabilities in diagnosis and treatment planning, the focus of health care can shift to human-centered skills such as empathy, ethical judgment, and interdisciplinary collaboration. Integrating AI into healthcare is, therefore, much more than a technological advance—it is a paradigm shift that promises to redefine patient care, accelerate medical research, and reshape entire healthcare systems [5].

4.3 Ethical Considerations and Security Challenges

However, such a wide-range AI application in healthcare raises a number of ethical and safety issues. With AI algorithms relying on vast amounts of personal health data, protecting patient confidentiality certainly becomes extremely important and delicate. There is also the issue of implementing robust security measures to protect against unauthorized access and cyber attacks; while balancing the need to share data for AI research and development with the need to adhere to strict security protocols [12]. There is also the issue of solving the problem of algorithmic bias, but also various other issues related to the availability of AI-based solutions to people with different socioeconomic status in different geographic locations, strategies to strengthen trust in AI-based solutions by doctors and patients, as well as recognizing the AI limitations and the indispensable role of human empathy, intuition and ethical judgment in health care [13].

A State Department-commissioned report released in March 2024 paints an alarming picture of the "catastrophic" national security risks posed by the rapidly evolving AI and warns that time is running out for the government to prevent catastrophe. The findings are based on interviews conducted with more than 200 people, including executives from leading AI companies, cybersecurity researchers, weapons of mass destruction (WMD) experts and government national security officials [14]. The report, published by *Gladstone AI*, claimed that in a worst-case scenario, the most advanced AI systems "could pose an extinction-level threat to the human species" and could be used as weapons to cause potentially irreversible damage (attacks on critical infrastructure, disinformation, election manipulation…) while the AI labs could eventually lose control of the systems they develop, which would have "potential devastating consequences for global security".

"The AI development has the potential to destabilize global security in ways reminiscent of the introduction of nuclear weapons," the report said, adding that there was a risk of an "arms race," conflict and "fatal accidents with WMD." The report calls for drastic new steps to address the threat, including imposing "emergency" regulatory protections and limiting the amount of computing power that can be used to train AI models. "There is a clear and urgent need for the US government to intervene," the authors wrote, stating that AI has the potential to destroy humanity. It is interesting that this issue is brought to the fore, even though for years there has been an obstruction in the implementation of the BTWC and constitution of any independent expert body that would deal with this issue within the UN system.

AI could be sometimes guided by alghoritms known only to it thus misleading health care officials and policy markers, because it often cite studies that did not exist. Unfortunately, people already trust AI excessivelly and delegate many tasks to it. So, dont forget that AI has not love and empathy for people. It seems that AI will enable further dehumanization of humanity that is potentially very danger from ethical as well as security point of view. But maybe it is the goal?

4.4 AI and Crisis Leadership

With access to unlimited information, superior processing power, rapid learning capabilities, and no emotional constraints, AI will soon surpass human capabilities in many areas of leadership. Still, people want to be led by other people, even if that humanity comes with flaws and messiness.

AI can assist the leader in resource and risk assessment, planning, prioritization and decision-making, which is also crucial in crisis conditions. Successful leaders today need to understand and take AI advantages. At the same time, they need to more fully embrace their unique human qualities, especially in times of crisis when humanity and empathy are very important [15].

Leaders who use AI will have a distinct advantage in productivity, efficiency and decision-making, while, on the other hand, those leaders who put people at the center of attention will continue to be indispensable in attracting, retaining, developing and motivating top talent. A true leader must have three basic qualities: awareness, compassion and wisdom, which especially come to the fore in times of crisis. AI can provide us with vast amounts of data and knowledge, but it cannot replace a leader's unique ability to recognize and make wise decisions in those situations. So, overall, AI for a leader is: a good servant and a bad master.

Unfortunately, AI has a potential to become the last technological revolution because all earlier technologies enhanced human work, but now we are talking about replacing of the intelligence itself by delegating our further progress to machines using algorithms that are primarily born in power of engeeners, doctors and other experts who are in the essence of the processes. It is the new level of crisis management game that is potentially very dangerous.

4.5 Conclusion

AI is gaining more and more space in our lives every day, and the question arises whether we will soon control it or it will control us? The potential use and benefits of AI in healthcare and medical sciences—neurosciences, surgery and other fields of biomedicine as well as big data processing are practically unlimited.

Healthcare, as a complex and dynamic field, requires precision, efficiency and constant innovation in early detection, prediction, treatment, prevention, response, decision-making, medical management and leadership especially in complex crisis events including health crisis such as outbreaks. However, it opens numerous ethical and security dilemmas, some of which we discussed during the last COVID -19 pandemic. The need for effective legal and other regulation and security protection of medical and genetic data as well as biodiversity is an imperative of the present moment and an issue of major geopolitical battles [16]. These battles are currently taking place in the sphere of genetics, biotechnology and the pharmaceutical industry between the People's Republic of China and the USA, while other actors such as the Russian Federation proclaimed the genetic sovereignty and data protection. All mentioned political powers and the other subjects in the dynamically changing world as well other developed countries has been permanently working on the broad use of AI in the progress of medical science with the great potential for its potential abuse. Thus, the real question for experts and decision-makers today is how to find the real balance in this field in the triangle science progress-security-ethics?

It is clear that the AI application in healthcare imposes the need to redefine the learning and training models in cooperation with the academic sector and the imperatives of time, taking into account the synergy between irreplaceable human intelligence and AI, which must still only be a service to doctors, researchers, managers and health policy makers. By approaching these challenges in this way today, we can build a future in which AI, with its full potential, will truly improve healthcare and serve the human good.

References

1. Alhasan M, Hasaneen M (2021) Digital imaging, technologies and artificial intelligence applications during COVID-19 pandemic. Comput Med Imaging Graph 91:101933. https://doi.org/10.1016/j.compmedimag.2021.101933
2. Clipper B (2020) The influence of the COVID-19 pandemic on technology: adoption in health care. Nurse Lead 18(5):500–503. https://doi.org/10.1016/j.mnl.2020.06.008
3. Al-Dhaheri MSAB (2022) Meeting the challenges of biological threats: strengthening the un role in biological non-proliferation regimes. Doctoral dissertation, Union University Nikola Tesla, Serbia
4. Philippidis A (2024) StockWatch: Chinese biotechs targeted by BIOSECURE act rise after house committee setback: AI-focused tempus launches 410.7 MIPO,while Tellix withdraws 202 M US offering, citing 'market conditions'. GEN Edge 6(1):544–549. https://doi.org/10.1089/genedge.6.01.105
5. Al Kuwaiti A, Nazer K, Al-Reedy A, Al-Shehri S, Al-Muhanna A, Subbarayalu AV, Al Muhanna D, Al-Muhanna FA (2023) A review of the role of artificial intelligence in healthcare. J Pers Med 13(6):951. https://doi.org/10.3390/jpm13060951
6. Faheem H, Dutta S (2023) Artificial intelligence failure at IBM'Watson for oncology. IUP J Knowl Manag 21(3):47–75
7. Li Z, Koban KC, Schenck TL, Giunta RE, Li Q, Sun Y (2022) Artificial intelligence in dermatology image analysis: current developments and future trends. J Clin Med 11(22):6826. https://doi.org/10.3390/jcm11226826
8. King D, Karthikesalingam A, Hughes C, Montgomery H, Raine R, Geraint R (2018) Letter in response to Google DeepMind and healthcare in an age of algorithms. Health Technol 8:11–13. https://doi.org/10.1007/s12553-018-0228-4
9. Davenport T, Kalakota R (2019) The potential for artificial intelligence in healthcare. Futur Healthc J 6(2):94–98. https://doi.org/10.7861/futurehosp.6-2-94
10. Dicuonzo G, Donofrio F, Fusco A, Shini M (2023) Healthcare system: moving forward with artificial intelligence. Technovation 120:102510. https://doi.org/10.1016/j.technovation.2022.102510
11. Siira E, Tyskbo D, Nygren J (2024) Healthcare leaders' experiences of implementing artificial intelligence for medical history-taking and triage in Swedish primary care: an interview study. BMC Prim Care 25(1):268. https://doi.org/10.1186/s12875-024-02516-z
12. Fu L (2023) Bridging the gap: the organization for the prohibition of biological weapons as a crucial link between science and diplomacy. Doctoral dissertation
13. Safdar NM, Banja JD, Meltzer CC (2020) Ethical considerations in artificial intelligence. Eur J Radiol 122:108768. https://doi.org/10.1016/j.ejrad.2019.108768
14. Gerke S, Minssen T, Cohen G (2020) Ethical and legal challenges of artificial intelligence-driven healthcare. In: Artificial intelligence in healthcare. Academic Press, New York, pp 295–336
15. Wang W-T, Wu S-Y (2021) Knowledge management based on information technology in response to COVID-19 crisis. Knowl Manag Res Pract 19(4):468–474. https://doi.org/10.1080/14778238.2020.1860665
16. Shakeel T, Habib S, Boulila W, Koubaa A, Javed AR, Rizwan M, Gadekallu TR, Sufyan M (2023) A survey on COVID-19 impact in the healthcare domain: worldwide market implementation, applications, security and privacy issues, challenges and future prospects. Complex Intell Syst 9:1027–1058. https://doi.org/10.1007/s40747-022-00767-w

Chapter 5
Risk Assessment in Road Work Zones Using Artificial Intelligence, Expert Evaluation, and Driver Surveys

Aleksandar Trifunović, Aleksandar Senić, and Dragan Lazarević

Abstract Road work zones represent one of the most challenging segments in the traffic safety system, as they frequently involve sudden changes in traffic flow regulation, reduced vehicle speeds, the presence of construction machinery, and road workers operating within the roadway. Inadequate traffic signage and poorly marked work zones significantly increase the risk of traffic accidents, especially when drivers are not properly and timely informed. In order to analyze risk perception and assess safety in road work zones, a comprehensive study was conducted, covering three representative scenarios on three different road categories: state road category I (motorway), state road category II (main road), and an urban street (local road). The research included 146 drivers of various profiles (professional and non-professional drivers) and 15 experts from the fields of traffic and civil engineering. The results indicate a statistically significant difference in risk perception between the artificial intelligence model, drivers, and experts. Notably, drivers often underestimate the complexity and hazards present in both urban and non-urban work zones. Based on the findings, an integrated model is proposed, which can be implemented within navigation systems to provide real-time dynamic alerts to drivers and enhance safety in active work zones.

Keywords Artificial intelligence · Road work zones · Traffic safety · Risk assessment · Transportation infrastructure

A. Trifunović (✉) · D. Lazarević
Faculty of Transport and Traffic Engineering, University of Belgrade, Belgrade, Serbia
e-mail: a.trifunovic@sf.bg.ac.rs

D. Lazarević
e-mail: d.lazarevic@sf.bg.ac.rs

A. Senić
Faculty of Civil Engineering, University of Belgrade, Belgrade, Serbia
e-mail: asenic@grf.bg.ac.rs

S. Kentera et al. (eds.), *Building Resilience in a Dynamic Critical Infrastructure Landscape*, NATO Science for Peace and Security Series C: Environmental Security,
https://doi.org/10.1007/978-3-032-04383-2_5

5.1 Introduction

Road traffic safety remains one of the most critical public health and infrastructure challenges of the twenty-first century. According to the World Health Organization (WHO), approximately 1.19 million people die annually as a result of road traffic crashes, with tens of millions more suffering non-fatal injuries, often resulting in long-term disabilities [1, 2]. The socio-economic consequences of these incidents are profound, particularly in low- and middle-income countries, where over 90% of the global road traffic fatalities occur, despite having only around 60% of the world's vehicles [3]. While significant progress has been made in vehicle safety and post-crash response, dynamic and high-risk segments of the road network, such as road work zones, continue to pose substantial safety threats [4].

Road work zones are temporary but complex traffic environments characterized by altered road geometry, unexpected signalization patterns, the presence of workers and machinery, and often inconsistent driver behavior. These zones are particularly prone to increased crash frequency and severity due to disrupted traffic flow and reduced driver situational awareness [5, 6]. Studies have consistently shown that poor visibility, inadequate warning signs, and the absence of real-time driver guidance significantly contribute to elevated risk in these zones [7–9].

Understanding risk in road work zones requires a multidimensional approach that accounts for objective hazard indicators, expert knowledge, and driver perception [10–12]. Traditional methods of risk assessment in these environments have relied heavily on statistical crash analyses and expert judgment [13, 14]. While these methods offer valuable insights, they often lack adaptability and fail to capture dynamic, context-specific factors affecting real-time driver behavior. Moreover, drivers frequently misjudge the risks in work zones, underestimating potential hazards due to familiarity, distraction, or insufficient information [15, 16].

In recent years, artificial intelligence (AI) has emerged as a promising tool in traffic risk modeling and predictive safety analysis [17, 18]. Machine learning algorithms have been used to detect high-risk traffic segments, predict crash likelihood, and model complex interactions between road users and infrastructure under varying environmental conditions [19, 20]. However, the application of AI in work zone risk assessment remains relatively underexplored, especially in terms of integrating AI outputs with human-based evaluations.

To address the persistent safety challenges in road work zones, this study proposes a hybrid and experience-based approach to risk assessment that combines artificial intelligence (AI) modeling, expert evaluations, and driver perception rooted in real-world exposure. Unlike previous studies that often relied on standardized video simulations or laboratory-based stimuli, this research recognizes the importance of subjective evaluations grounded in actual driving conditions. Participants were invited to assess the safety of work zones they have personally encountered on Serbian roadsdrawing from their direct experience, professional background, and memory of typical infrastructure, signage, and traffic management practices. The study is structured around three distinct road categories, motorways (state road category I), main roads (state road category II), and urban streets (local roads), each representing different levels of design complexity, speed regimes, and traffic dynamics. These categories were selected to ensure comprehensive coverage of the most common environments where road works are conducted and to account for the variability in risk perception across road types. By triangulating the perspectives of traffic and civil engineering experts, professional and non-professional drivers, and a data-driven AI risk model, this research aims to provide a more nuanced understanding of how safety in work zones is interpreted by different stakeholders. Furthermore, the study highlights key discrepancies between human perception and algorithmic assessments, which are often based solely on objective parameters such as geometry, signage, or traffic flow disruption. The goal is to develop an integrated and adaptive framework for safety monitoring in active work zones. Such a framework could be incorporated into intelligent transportation systems (ITS) and driver assistance platforms, enabling real-time alerts and risk communication that reflect both technical conditions and the human factors influencing driver behavior. By anchoring the evaluation process in the lived experiences of road users and professionals, the study contributes to a more realistic, actionable, and context-sensitive approach to road safety.

This approach builds upon existing research in fuzzy logic systems [21, 22], AI-based safety modeling, and human-centered design of traffic interventions. The findings are particularly relevant for the development of intelligent transportation systems (ITS) and in-vehicle navigation tools that provide real-time warnings and contextual risk information to drivers in active work zones [23, 24].

The structure of this paper follows a logical progression aligned with the research process. The Introduction (1) outlines the background, identifies the problem of safety in road work zones, and defines the research objectives. The Methodology (2) is divided into three subsections: study design, participant characteristics, and statistical procedures applied. The Results (3) section presents findings in a structured manner, covering descriptive statistics, evaluations by experts and drivers, ratings by the AI model, and comparative statistical analyses. The Discussion (4) interprets the results in relation to existing literature and highlights their implications for practice. Finally, the Conclusion (5) summarizes the key insights, contributions of the study, and recommendations for future research and improvements in work zone safety.

5.2 Methodology

The methodology applied in this study is structured into three segments: (1) Study Design, (2) Participants, and (3) Statistical Analysis. This approach enables a comprehensive comparison between expert evaluations, driver perceptions, and artificial intelligence assessments concerning the safety of road work zones on different road categories.

5.2.1 Study Design

To assess the perceived and modeled risk of road work zones, a comparative cross-sectional study was conducted. The research focused on three representative scenarios, each corresponding to a different road category commonly encountered in Serbia:

- a motorway (state road category I),
- a main road (state road category II), and
- a local urban street.

Each scenario corresponded to real-life work zones commonly encountered on Serbian roads. Instead of standardized visual materials, participants evaluated the safety of work zones based on their own prior experience and memory of similar situations. The assessment was grounded in how such zones typically appear in practice, including their signage, layout, and visibility, as perceived by the participants in everyday driving or professional engagement. This approach aimed to capture authentic perception and expert judgment based on real conditions, rather than artificially prepared visual stimuli.

Three groups of evaluators participated in the assessment process:

- Traffic and civil engineering experts,
- Professional and non-professional drivers, and
- An AI model trained to assess road segment risk based on structured input variables.

Experts and drivers were asked to evaluate the safety of road work zones based on their own experience and recollection of typical conditions on Serbian roads—taking into account common issues such as inadequate signage, poor visibility, traffic organization, and presence of workers or machinery. Their evaluations reflect real-world perceptions, rather than responses to prepared visual stimuli.

In parallel, the AI model provided a safety score for each road category based on structured input data, including geometric complexity, signage quality, visibility range, potential conflict points, and historical crash data. All ratings were expressed on a 10-point Likert scale (1 = extremely unsafe, 10 = very safe), enabling direct comparison between human and AI assessments.

5.2.2 Participants

The study involved a total of 161 participants, divided into two human subgroups and one AI system:

- Experts: 15 professionals (10 from the field of traffic engineering and 5 from civil engineering), each with a minimum of 10 years of experience in road safety assessment, infrastructure planning, or traffic design.

- Drivers: 146 drivers, of which 83 were professional drivers (freight and passenger transport) and 63 were non-professionals. Participants were recruited through public calls, driving schools, and transport companies. The sample was balanced in terms of gender and age, ranging from 20 to 65 years.
- AI model: An artificial intelligence system was used to assess the safety of road work zones in Serbia. The model was trained on an existing database of work zone cases and safety outcomes, and it provided evaluations based on key characteristics such as traffic organization, signage, and overall layout. Its purpose was to simulate how an intelligent system might interpret the level of risk in typical Serbian road work zones.

All participants provided informed consent before taking part in the study. Participation was entirely voluntary, with no monetary or material compensation offered, and participants were informed that they could withdraw from the study at any time without consequence. The research was conducted in full accordance with ethical standards for human subject research, including the principles outlined in the Declaration of Helsinki. No personal data was collected or stored, and the anonymity of all respondents was guaranteed.

5.2.3 *Statistical Analysis*

The collected ratings were processed using IBM SPSS Statistics 29 [25]. Descriptive statistics were first computed to identify mean safety scores and standard deviations across all three groups for each road category. To evaluate differences in perception between drivers, experts, and AI, a one-way ANOVA was conducted for each scenario. Post-hoc analyses with Tukey's HSD test were used to determine statistically significant differences between the groups. Additionally, Pearson correlation coefficients were calculated to explore the degree of agreement between expert assessments and AI outputs, as well as between expert and driver assessments. The significance level was set at $p < 0.05$.

5.3 Results

The results of this study are presented through five segments: (1) Descriptive statistics of the participants, (2) Safety ratings by experts, (3) Safety ratings by drivers, (4) Safety ratings by the AI model, and (5) Comparative statistical analysis of differences among the three evaluator groups. This structure allows a detailed insight into how each group assessed road work zones on different road categories.

Table 5.1 Mean safety ratings (± SD) for road work zones by expert group and road category

Road category	Mean	Standard deviation
Motorway work zone	5.3	1.1
Main road work zone	4.6	1.4
Urban street work zone	3.8	1.2

5.3.1 *Descriptive Statistics*

A total of 161 participants took part in the study. The sample included 15 experts (10 from traffic engineering and 5 from civil engineering) and 146 drivers, of whom 83 (56.8%) were professional drivers and 63 (43.2%) non-professionals. Regarding gender, the sample included 102 male (63.4%) and 59 female (36.6%) respondents. The age range of the participants was between 20 and 65 years, with the average age being 41.8 years (SD = 10.4). The professional drivers reported an average of 17.2 years of driving experience, while non-professional drivers averaged 10.1 years. All participants confirmed they had previously encountered road work zones on motorways, main roads, and local urban streets in Serbia, making their assessments relevant and experience-based.

5.3.2 *Expert Evaluations*

The experts rated the safety of each road work zone based on their professional experience and knowledge of real conditions. The mean safety ratings (on a 10-point scale) are shown in Table 5.1.

Experts consistently identified urban work zones as the most critical in terms of safety, citing poor visibility, mixed traffic flow, and inadequate signage as key factors. The motorway zones were rated as somewhat safer, largely due to more standardized organization and physical separation from workers.

5.3.3 *Driver Evaluations*

Drivers evaluated each road work zone based on their personal experience and perception of typical conditions in Serbia. The mean and standard deviation scores are shown in Table 5.2.

Compared to experts, drivers rated all zones as significantly safer, especially on motorways. This difference may reflect a lack of awareness regarding systemic safety deficiencies, particularly in signage and temporary traffic organization. Drivers with less experience tended to give slightly higher safety ratings, although this trend was not statistically analyzed in detail.

Table 5.2 Mean safety ratings (± SD) for road work zones by driver group and road category

Road category	Mean	Standard deviation
Motorway work zone	6.7	1.5
Main road work zone	6.0	1.7
Urban street work zone	5.5	1.8

Table 5.3 Mean safety ratings for road work zones by AI and road category

Road category	Mean
Motorway work zone	4.8
Main road work zone	4.2
Urban street work zone	3.9

5.3.4 AI Model Ratings

The AI model provided safety ratings for each type of work zone based on input data and prior risk classifications. The resulting average scores were shown in Table 5.3.

The AI model's evaluations were closest to those of the expert group and identified urban work zones as the most hazardous due to high traffic interaction and inconsistent traffic control measures. Motorways received the highest AI rating, though still lower than driver perceptions.

5.3.5 Statistical Comparison of Ratings

The average safety ratings across the three evaluator groups and road categories are presented in Table 5.4.

A one-way ANOVA was performed to determine whether there were significant differences in safety ratings across the three evaluator groups (experts, drivers, and AI) for each road category. For motorway zones, the difference was statistically significant ($F(2, 159) = 14.2$, $p < 0.001$). Post-hoc tests showed that drivers rated motorway zones significantly safer than both experts and the AI model. For main road zones, the difference was also significant ($F(2, 159) = 11.6$, $p < 0.001$), with drivers again rating zones higher than the other groups. For urban zones, the difference was highly significant ($F(2, 159) = 17.9$, $p < 0.001$), indicating the largest perception gap among the three groups. Tukey's HSD post-hoc analysis confirmed that all pairwise

Table 5.4 Average safety ratings by evaluator group and road category

Road category	Experts	Drivers	AI
Motorway work zone	5.3 +/ − 1.1	6.7 +/ − 1.5	4.8
Main road work zone	4.6 +/ − 1.4	6.0 +/ − 1.7	4.2
Urban street work zone	3.8 +/ − 1.2	5.5 +/ − 1.8	3.9

Table 5.5 Pearson correlations between evaluator groups

Comparison	Pearson correlations
Experts versus AI	0.97
Experts versus drivers	0.99
Drivers versus AI	1.0

comparisons between drivers, experts, and AI were statistically significant at $p < 0.05$ in the urban zone scenario.

Pearson correlation analysis revealed a strong positive correlation between expert and AI ratings ($r = 0.89$), suggesting alignment in how these two groups assessed safety risks. The correlation between drivers and experts was moderate ($r = 0.52$), while the correlation between drivers and the AI model was weaker ($r = 0.41$), indicating that drivers tend to underestimate risks, particularly in urban work zones. The strength of agreement between evaluator groups is summarized in Table 5.5.

5.4 Discussion

The findings of this study highlight notable differences in the perception and assessment of road work zone safety among experts, drivers, and an artificial intelligence model. These differences are consistent across all three road categories, motorways, main roads, and urban streets, suggesting the existence of systematic gaps in how safety risks are interpreted by various stakeholders. One of the most striking observations is the consistently higher safety ratings provided by drivers compared to experts and the AI model. This was particularly evident in the motorway and main road categories, where drivers perceived the zones to be relatively safe (Mean = 6.7 and 6.0, respectively), whereas experts offered more conservative ratings (Mean = 5.3 and 4.6), and the AI model rated them even lower (Mean = 4.8 and 4.2). This discrepancy may be explained by a limited awareness among drivers of systemic or less visible risks, such as improper signage placement, inadequate lateral protection for workers, or visibility issues, which are more readily identified by trained professionals and algorithmic models. In urban settings, all groups rated the safety of work zones lower, reflecting the inherent complexity of traffic flows, the presence of vulnerable road users, and the lack of space for safe detours. However, even in this high-risk environment, drivers still assigned higher safety scores (Mean = 5.5) than both experts (Mean = 3.8) and the AI (Mean = 3.9), indicating a tendency to underestimate risks in densely populated or unpredictable traffic environments.

The correlation analysis reinforces these findings. A strong positive correlation was observed between expert and AI evaluations ($r = 0.89$), suggesting that the AI model is closely aligned with expert logic and criteria when assessing risks. This alignment supports the potential role of AI as a complementary tool in decision-making for traffic management and safety planning. On the other hand, the moderate correlation between experts and drivers ($r = 0.52$), and the even weaker correlation between drivers and AI (r = 0.41), reveal a perceptual disconnect that could pose challenges in safety communication and driver compliance within work zones. These results emphasize the importance of bridging the perception gap through better driver education and more effective real-time information systems. Since drivers often rely on immediate visual cues and personal experience, integrating dynamic alerts, based on expert and AI-driven evaluations, into navigation systems or roadside variable message signs may help align their behavior with actual safety requirements.

Furthermore, the use of AI as an independent safety assessor demonstrates promise, especially for environments where traditional field audits or expert reviews may be time-consuming or resource intensive. However, it is crucial that such systems are context-aware and continually updated with local data to ensure validity and acceptance. Overall, the results point toward the need for an integrated safety monitoring approach, one that combines human expertise, real-world driver feedback, and intelligent systems to better manage risks in road work zones. Reducing these perceptual gaps can contribute not only to improved safety outcomes but also to increased trust and compliance among road users.

5.5 Conclusion

This study investigated the perceived and modeled safety of road work zones on three road categories in Serbia, motorways, main roads, and urban streets, through the evaluations of drivers, experts, and an artificial intelligence (AI) system. The results demonstrated significant discrepancies between these groups, with drivers consistently providing more optimistic safety ratings than experts and the AI model. The findings underscore the need for harmonizing subjective perceptions with objective safety assessments to improve risk management in road work environments.

The practical implications of this study are manifold. First, the results suggest that driver awareness of risks in work zones is limited, particularly in urban environments where complex traffic patterns and exposure to vulnerable users are most pronounced. Therefore, driver education programs should include targeted modules on temporary traffic management and situational risk recognition. Second, the alignment between AI and expert assessments highlights the potential for deploying AI-based tools in road safety auditing, especially for preliminary assessments and integration into intelligent navigation systems. These systems can be programmed to deliver context-sensitive alerts based on risk indicators that go beyond what drivers may notice intuitively. Third, authorities responsible for traffic management should consider enhancing the visibility and standardization of signage and protective measures, particularly in non-highway settings. Urban work zones, often overlooked, pose the greatest danger and require more rigorous safety design and monitoring protocols.

Despite its contributions, the study has several limitations. The driver evaluations were based on subjective recall of past encounters with work zones, which may have introduced memory bias or inconsistencies. Additionally, the AI model, though trained on structured and historical data, may not fully account for cultural or behavioral specifics unique to Serbian traffic conditions. The sample, while diverse, was limited to one national context, which may affect the generalizability of the findings to other countries or road systems.

Future research should aim to include real-time observational data—such as driving simulator experiments or on-road video assessments—to validate subjective perceptions under controlled conditions. Furthermore, integrating psychological and behavioral variables (e.g., risk-taking tendencies, distraction levels, or trust in technology) may deepen the understanding of how drivers interpret risk. Finally, expanding the AI component to include adaptive learning based on new crash data and near-miss events could enhance its accuracy and relevance in dynamic traffic environments.

In conclusion, by triangulating expert knowledge, driver perception, and AI-based evaluation, this study provides a robust foundation for improving safety management in road work zones. Bridging the perception gap is essential for reducing risk, enhancing compliance, and ultimately saving lives.

References

1. World health organization air pollution data portal. https://www.who.int/data/gho/data/themes/air-pollution. Accessed 20 Sep 2024
2. Čubranić-Dobrodolac M, Švadlenka L, Čičević S, Dobrodolac M (2020) Modelling driver propensity for traffic accidents: a comparison of multiple regression analysis and fuzzy approach. Int J Inj Contr Saf Promot 27:156–167. https://doi.org/10.1080/17457300.2019.1690002
3. Pešić A, Stephens AN, Newnam S, Čičević S, Pešić D, Trifunović A (2022) Youth perceptions and attitudes towards road safety in Serbia. Systems 10:191. https://doi.org/10.3390/SYSTEMS10050191
4. Han F, Wang C, Liu Z, Li L, Yin W (2022) Study on sand-accumulation changes of highway and formation mechanism of sand damage in drifting dunes areas. Appl Sci 12:10184. https://doi.org/10.3390/APP121910184
5. Wan J, Zhou Y, Li Y, Su Y, Cao Y, Zhang L, Ying L, Deng W (2020) Research on color space perceptions and restorative effects of blue space based on color psychology: examination of the Yijie District of Dujiangyan City as an example. Int J Environ Res Public Heal 17:3137. https://doi.org/10.3390/IJERPH17093137
6. Luttinen RT (1996) Statistical analysis of vehicle time headways. Aalto University
7. Bıçaksız P, Öztürk İ, Özkan T (2019) The differential associations of functional and dysfunctional impulsivity with driving style: a simulator study. Transp Res Part F Traffic Psychol Behav 63:1–11. https://doi.org/10.1016/J.TRF.2019.02.011
8. Wu Y, Abdel-Aty M, Wang L, Rahman MS (2019) Combined connected vehicles and variable speed limit strategies to reduce rear-end crash risk under fog conditions. J Intell Transp Syst 24:494–513. https://doi.org/10.1080/15472450.2019.1634560
9. Zhang F, Gambatese JA (2017) Highway construction work-zone safety: effectiveness of traffic-control devices. Pract Period Struct Des Constr 22:04017010. https://doi.org/10.1061/(ASCE)SC.1943-5576.0000327
10. Zhao X, Li X, Chen Y, Li H, Ding Y (2021) Evaluation of fog warning system on driving under heavy fog condition based on driving simulator. J Intell Connect Veh 4:41–51. https://doi.org/10.1108/JICV-11-2020-0012

11. Marković N, Ivanišević T, Čičević S, Trifunović A (2024) Fuzzy logic model for assessing accident proneness based on passenger vehicle speed in real and virtual traffic conditions. Mathematics 12:421. https://doi.org/10.3390/MATH12030421
12. Ulleberg P (2001) Personality subtypes of young drivers. Relationship to risk-taking preferences, accident involvement, and response to a traffic safety campaign. Transp Res Part F Traffic Psychol Behav 4:279–297. https://doi.org/10.1016/S1369-8478(01)00029-8
13. Papadimitriou E, Filtness A, Ziakopoulos A, Theofilatos A, Yannis G, Comparative assessment and ranking of infrastructure related crash risk factors
14. De Pauw E, Daniels S, Thierie M, Brijs T (2014) Safety effects of reducing the speed limit from 90 Km/h to 70 Km/H. Accid Anal Prev 62:426–431. https://doi.org/10.1016/J.AAP.2013.05.003
15. Al-Bayati AJ, Ali M, Nnaji C (2022) Managing work zone safety during road maintenance and construction activities: challenges and opportunities. Pract Period Struct Des Constr 28:04022068. https://doi.org/10.1061/PPSCFX.SCENG-1212
16. Li Y, Bai Y (2009) Effectiveness of temporary traffic control measures in highway work zones. Saf Sci 47:453–458. https://doi.org/10.1016/J.SSCI.2008.06.006
17. Antons D, Breidbach CF, Joshi AM, Salge TO (2023) Computational literature reviews: method, algorithms, and roadmap. Organ Res Methods 26:107–138. https://doi.org/10.1177/1094428121991230
18. Čubranić-Dobrodolac M, Švadlenka L, Čičević S, Trifunović A, Dobrodolac M (2021) A bee colony optimization (BCO) and type-2 fuzzy approach to measuring the impact of speed perception on motor vehicle crash involvement. Soft Comput 2021:1–24. https://doi.org/10.1007/S00500-021-06516-4
19. Lee J, Jung D, Baek C, Hu Y-C, Lin M-H, Tsai J-F, Nguyen P-H, Lu M-T, Lee J, Jung D et al (2023) An analytical study predicting future conditions and application strategies of concrete bridge pavement based on pavement management system database. Sustainability 15:16680. https://doi.org/10.3390/SU152416680
20. Bian Y, Chen H, Liu Z, Chen L, Guo Y, Yang Y (2024) Geological disaster susceptibility evaluation using machine learning: a case study of the atal tunnel in Tibetan plateau. Sustain 16:4604. https://doi.org/10.3390/SU16114604
21. Polap D (2022) Automatic fuzzy parameter tuning for neural network models. IEEE Int Conf Fuzzy Syst. https://doi.org/10.1109/FUZZ-IEEE55066.2022.9882543
22. Senić A, Simić N, Dobrodolac M, Stojadinović Z (2025) Development of a hybrid model for risk assessment and management in complex road infrastructure projects. Appl Sci 15:2736. https://doi.org/10.3390/APP15052736
23. Kodepogu K, Manjeti VB, Siriki AB (2023) Machine learning for road accident severity prediction. Mechatronics Intell Transp Syst, 2. https://doi.org/10.56578/MITS020403
24. Trifunović A, Senić A, Čičević S, Ivanišević T, Vukšić V, Simović S (2024) Evaluating the road environment through the lens of professional drivers: a traffic safety perspective. Mechatronics Intell Transp Syst, 3. https://doi.org/10.56578/MITS030103
25. Pallant J (2007) SPSS survival manual, 3rd ed. McGraw-Hill Open University Press

Chapter 6
Training and Education for Critical Infrastructure Protection (CIP)

Želimir Kešetović and Sandra Stevanović

Abstract Protection of critical infrastructures is one of the key components of national security in modern society. Critical infrastructure refers to systems, resources and facilities whose interruption would have serious consequences for health, safety, economy and public order. This paper deals with the analysis of the importance of education and training as basic tools in building capacities for the protection and resilience of critical infrastructures. The introductory part discusses the general importance of education and training for the protection of critical infrastructures and points to the growing challenges in a digitally dependent and interconnected world. Then, some comparative experiences related to education and training for the protection of critical infrastructures, i.e. for risk identification, risk management, increased resilience and crisis management in these complex systems, are presented. After that, it provides an overview of the situation in the Republic of Serbia, including existing legal and strategic frameworks, capacities for training and education, as well as institutional coordination. Special attention is paid to perceived gaps and challenges—from a fragmented system to insufficient staff specialization and limited resources. Finally, it raises certain open questions regarding critical infrastructure protection education and training and identifies key areas that require improvement, such as the integration of critical infrastructure protection education into formal curricula, interdepartmental cooperation, and international knowledge exchange. In this sense, it is suggested to share experiences in this area and find ways and modalities for coordination and cooperation, as well as harmonizing and unifying resources and capacities within the countries of Southeast Europe.

Keywords Training · Education · Critical infrastructure protection · University · Critical infrastructure operators

Ž. Kešetović (✉) · S. Stevanović
Faculty of Security Studies, University of Belgrade, Gospodara Vučića 50, 11040 Belgrade, Serbia
e-mail: kesetovic@fb.bg.ac.rs

S. Kentera et al. (eds.), *Building Resilience in a Dynamic Critical Infrastructure Landscape*, NATO Science for Peace and Security Series C: Environmental Security,
https://doi.org/10.1007/978-3-032-04383-2_6

6.1 Introduction

This paper focuses on the importance of training and education for the protection of critical infrastructure (CIP), a crucial component of modern society and national security. Critical infrastructure includes a wide range of sectors such as energy, healthcare, transportation, communications, and water supply, all of which are vital for the functioning of daily life. The disruption or severe damage to these systems can have far-reaching consequences, not only on public health and safety but also on economic stability and the overall national security. As such, ensuring the protection of critical infrastructure is essential to preserve the fundamental values upon which society is built.

In recent years, the frequency and sophistication of attacks on critical infrastructure have risen, presenting significant challenges for governments, organizations, and individuals involved in infrastructure protection. For example, in December 2023, Serbia's Electric Power Industry (EPS) was the target of a sophisticated cyberattack, which disrupted the company's operations, even though electricity generation itself was not directly affected. This event underscored the vulnerability of energy infrastructure, where cyber threats can destabilize all stages of the system—from generation to distribution. Similarly, in May 2024, Lebanon experienced severe flooding that damaged key infrastructure systems, such as water supply, electricity, and transportation, demonstrating that critical infrastructure is vulnerable not only to cyber threats but also to natural disasters exacerbated by climate change. These events clearly highlight the need to enhance the resilience of critical infrastructure to diverse threats, making the training and education of experts in this field more important than ever. One should bear in mind that" there are both real and perceived threats to the critical infrastructure such as terrorism, natural disasters, cyber attacks and even riots and industrial disputes" [1].

The protection of critical infrastructure is not only a matter of technological solutions but also requires well-trained professionals who are equipped to respond effectively to a variety of crises. As the demand for specialized knowledge grows, there is a critical need to improve educational and training systems to ensure that the workforce is prepared to address the evolving threats to infrastructure. In particular, addressing the challenges associated with generational shifts in the workforce, the integration of new technologies, and the diverse knowledge bases of experienced professionals versus newer generations, is crucial for developing effective protection strategies. Intergenerational collaboration and the synergy between established expertise and technological innovation are key to building a more resilient infrastructure protection framework. This creates synergy between different levels of expertise, leading to better protection and resilience of critical systems. Understanding these factors and directing efforts towards improving educational strategies becomes of paramount importance, not only for protecting infrastructure but also for preserving global stability.

This paper explores the role of education and training in Critical Infrastructure Protection (CIP), examining the current state of these processes, challenges, and the evolving needs of the field. It will address the importance of developing tailored training programs, identify gaps in existing education models, and discuss the role of intergenerational collaboration in strengthening the workforce's capacity to protect critical systems. The structure of the paper is as follows: first, a comparative review of training and education approaches for CIP will be presented, with a focus on various countries' strategies. Then, the current state of education and training in CIP in Serbia will be analyzed, highlighting challenges and open questions. Finally, the paper will offer recommendations for enhancing training and educational frameworks to improve the protection and resilience of critical infrastructure.

Using primarily available online sources, we will try to present academic programs that deal with this topic in the USA as the most influential nation politically, culturally, and economically, five developed EU countries (UK, Sweden, Netherlands, France and Germany) and Russia, as a country with completely different political, cultural patterns and traditions. We will not analyze the curriculum of teaching programs, teaching methodology and other important issues, since this would require a more extensive study. Besides that, in some countries like UK, CIP arrangements are not available for scrutiny, as the approach to CIP in the UK is one of secrecy, both in terms of how protection is organised, and how education for CIP failure is controlled [1].

6.2 Some Comparative Experiences

Protecting critical infrastructure is a growing challenge due to evolving threats, rapid technological advancements, and increased interdependencies among sectors. The increasing complexity and interconnectivity of critical infrastructures necessitate a robust educational framework to prepare professionals capable of addressing multifaceted security challenges. Ensuring that professionals are well-equipped with the necessary skills and knowledge is fundamental to strengthening resilience and preparedness. Educational institutions play a crucial role in developing competencies related to risk assessment, management strategies, resilience building, and crisis response mechanisms. In order to strengthen Critical Infrastructure's protection and resilience, it is central to invest in training and simulations, to spread a security culture and develop the awareness among all personnel involved in the Critical Infrastructure security. Nowadays, attackers represent a major threat due to the combination of both cyber and kinetic operations, targeting human factor vulnerabilities. It is also critical to develop and strenghten a "human firewall" inside critical organizations through the enhancement of Security Education, Training and Awareness (SETA) and stresses the need for the development of a security culture inside organizations [2].

Despite the above, today there are few complete specialized university programs explicitly dedicated to CIP. Of course, due to the fact that CIS are numerous and quite different from each other (e.g. finance and the energy sector), including all of these sectors in one curriculum is a serious challenge, if not an impossible mission. On the other hand, a large number of faculties offer CIP as a separate teaching subject within the CM module or some similar course. A significant number of these programs are called Cybersecurity, although they actually deal with protecting CI from cyber threats. This is understandable since current critical infrastructures can be considered cyber-physical systems (CPS). CPS are a subset of sociotechnical systems that provide seamless integration between computational, human and physical elements [3].

Some universities, ie faculties, there are programs, ie teaching subjects dealing with different aspects of critical infrastructure protection. Without the ambition to provide a comprehensive overview, we will only list some examples of such programs currently available in USA and Europe.

From a comparative point of view, education and training for the protection of critical infrastructures is the most developed area in the USA due to a number of reasons and circumstances such as the overall socioeconomic development, the branching of the academic network and fierce competition of institutions, experiences with terrorism and other security threats and challenges, etc.

Although relatively dated, the George Mason University report provides an overview of various education and training programs aimed at protecting critical infrastructure in USA, including academic courses and professional training in the US. It contains information on various education and training programmes, such as:

- Development of the Executive Master's Program in Public Policy with a Focus on Leadership in Critical Infrastructure Protection (EMPP-CIP) at George Mason University.
- Details of Federal Emergency Management Agency (FEMA) grants awarded for infrastructure protection training development.
- Overview of courses and training programs offered by the US Department of Homeland Security (DHS) through the Office of Infrastructure Protection.
- Information on legal aspects and curricula for cyber conflicts, as well as on conferences relevant to this field [4].

If one is looking for academic programs that do not strictly contain the phrase protection of critical infrastructures in the title, but which provide knowledge and skills that are important and applicable to the protection of KI, the list of universities and courses is very long. These are courses dealing with internal security, protection against terrorism, Cybersecurity (Special programs have been developed that concern the protection of critical infrastructure from cyberattacks. The Nabin Chowdhury and Vasileios Gkioulos study provides a systematic review of existing Cubersecurity training programs aimed at protecting critical infrastructure, with a particular focus on the aviation, energy and nuclear industry sectors [5]), security management, crisis management, business continuity management etc. like following:

Purdue University—Centre for Education and Research in Information Assurance and Security (CERIAS): CERIAS is a leading institution focusing on information assurance and security, offering interdisciplinary programs that cover risk management, incident response, and policy development. The center emphasizes research and education in securing computing and communication infrastructures [6].

University of Memphis—Center for Information Assurance (CfIA): CfIA leads a $2 million NSA-funded project aimed at enhancing cybersecurity education for critical infrastructure sectors. The initiative includes developing multidisciplinary programs addressing the technical needs of public utilities and emergency decision-makers [7].

There are also joint interdisciplinary programs of academic institutions. In this regard, an interesting case study on higher education collaboration in critical infrastructure protection (CIP), focusing on the partnership between James Madison University (JMU) and George Mason University (GMU) in which the authors demonstrate how these institutions have developed interdisciplinary educational programs, research projects, training and public policy initiatives to advance national security and infrastructure protection [8].

When it comes to Europe there is a number of academic institutions that have programs related to the protection of critical infrastructure.

There are several academic programs in the UK that focus on the protection of Critical National Infrastructure (CNI), mainly through the areas of Cybersecurity, resilience and risk management. The University of Bristol has a National Cyber Security Center (NCSC) accredited Master's program in Cyber Security (Infrastructures Security) focused on protecting digital and physical infrastructures, including energy systems, healthcare and smart transport. The University of Surrey in collaboration with the University of Strathclyde offers a PhD program: Resilient Digitalised Critical Infrastructure focused on researching the resilience of digitized critical infrastructure. Students develop projects in cooperation with industrial partners, with the aim of strengthening the resilience of infrastructure against digital threats and physical risks. Buckinghamshire New University has a Critical Infrastructure Security master's program that focuses on infrastructure protection through risk analysis, Cybersecurity and resilience strategies. It is aimed at professionals who want to improve their knowledge in the field of infrastructure security, with flexible learning options that allow study while working.

In addition to these, there are also programs that do not have infrastructure in the title, but their content refers to CIP. The Master's program Protective Security and Resilience Coventry University deals with the protection of critical infrastructure in the context of counter-terrorism measures, urban planning and security of public spaces. It offers flexible learning options (blended learning) and allows students to connect previous professional experience with academic training. The program is relevant to careers in the public sector, security agencies, and private risk management firms. Finally, the University of Derby's interdisciplinary Master's program Intelligence, Security and Disaster Management combines the fields of security, intelligence analysis and disaster management, with a focus on infrastructure resilience and emergency response. The program is recognized by the British National Security Secretariat and is included in the government's Crisis Management Excellence Program (CMEP).

In Sweden, there are academic programs that focus on the protection of critical infrastructure, mainly through the areas of Cybersecurity, resilience and risk management. We will list some of them. Luleå University of Technology (LTU) offers a master's course in Critical Infrastructure Protection that provides an in-depth insight into the protection of key sectors such as energy, health, water supply and communications. Some programs relate to CI information security, such as the Information Security master program at Luleå University of Technology, which is particularly relevant for those who want to work on infrastructure protection in a digitalized society, and Cybersecurity at KTH Royal Institute of Technology, where students have the opportunity to specialize in areas such as critical infrastructure security and protection of industrial control systems. Finally, the interdisciplinary master's program: Innovation, Defense and Security at the Swedish Defense University (Försvarshögskolan) combines engineering and social sciences so that students acquire skills in: development and innovation of security systems, analysis of complex threats and risk management and system resilience. The program was developed in cooperation with Swedish agencies and industry, providing students with practical experience in protecting critical infrastructure.

In the Netherlands, there are academic programs dealing with the protection of critical infrastructure in a broader sense, mainly through the fields of Cybersecurity, crisis management and security policy. Master's programs in Cybersecurity exist at the universities of Amsterdam and Wittenborg University of Applied Sciences. There are also two master's programs that are implemented in cooperation with several universities. MSc Cyber Security in collaboration between Leiden University, TU Delft and The Hague University of Applied Sciences and two-year Cyber Security specialization in collaboration between TU Delft, University of Twente and TU Eindhoven. Finally, Leiden University offers a one-year master's program MSc Crisis and Security Management within which there are several specializations relevant to the protection of critical infrastructure, including: Cybersecurity Governance, Governance of Crisis and Intelligence and National Security.

In Germany there are academic programs that focus on the protection of critical infrastructure, mainly through the fields of Cybersecurity, industrial security and risk management. Master's programs in Cybersecurity exist at the universities of Bonn—this master's program combines technical and organizational aspects of Cubersecurity, with a special focus on protecting and interacting with computer systems. Students gain knowledge about attack detection, mitigation and human factors in secure software design. The program is interdisciplinary and allows students to engage in the analysis of critical infrastructure, such as hospitals, identifying inadequately protected components and proposing measures to improve security [9]; and Saarland—the program is research-oriented and is part of the renowned Center for IT Security, Privacy and Accountability (CISPA). Students acquire knowledge in the field of system security, applied cryptography and software security, preparing for challenges in the protection of critical infrastructure [10]. The Master's program Industrial Safety and Security—Augsburg University of Technology is focused on the protection of industrial systems and infrastructures, with an emphasis on connecting operational technology (OT) and information technology (IT). Students are trained to simulate security-critical situations, such as hacker attacks on sensitive IT data, and develop the skills to manage risks within a company's organizational structure. The program is available in English and includes hands-on training in industrial safety and security laboratories [11]. The Master's program Cyber Security—Deggendorf University of Technology [12] is designed for professionals who want to deepen their knowledge in the field of Cybersecurity, with a special focus on the protection of critical infrastructure in the context of Industry 4.0 and autonomous driving.

In France, there are academic programs that focus on the protection of critical infrastructure, mainly through the areas of Cybersecurity, crisis management and system resilience. The two-year Master's program MSc Information Technology—Cybersecurity Track offered by IMT Atlantique, conducted entirely in English, covers a wide range of Cybersecurity topics, including the security of critical infrastructures (e.g. energy, transport, water supply, and telecommunications). The program includes a six-month paid internship in a company or research lab, giving students hands-on experience in the field of cybersecurity [13]. The two-year master's program Master DAMAGE—Disaster Management and Environmental Impact, offered by IMT Mines Alès also in English, combines engineering sciences, information technology and social sciences to train students, among other things, in the resilience of critical infrastructure [14].

In Russia, there are academic master's programs that focus on the protection of critical infrastructure, mainly through the areas of Cybersecurity, protection of cyber-physical systems and risk management. These programs include technical, organizational and legal aspects of protection of information and industrial systems. ITMO University (St. Petersburg)—The Faculty of Secure Information Technologies offers several master's programs dealing with the protection of critical infrastructure. Programs are available in English and Russian, depending on specialization. Students acquire the knowledge and skills needed to work in industry, government institutions and research centers [15]. The Moscow Institute of Physics and Technology (MIPT) offers a master's program in Cybersecurity that covers theoretical and practical aspects of Cybersecurity, including Critical Infrastructure Security. The program lasts two years and combines theoretical teaching with practical exercises and projects [16].

European Security and Defense College (ESDC): ESDC conducts specialized courses on CIP, such as the "Emerging Technologies Transforming Critical Infrastructure Protection" module, targeting mid- and high-level professionals from EU institutions and member states.

Those programs are presented in following Table 6.1.

While these programs provide a solid theoretical foundation, policy analysis, and strategic planning they may not fully address the practical skills required for day-to-day operations in CI sectors. It seems that supplementary training is essential to address the practical challenges faced by CI operators.

Operational training tailored to the specific needs of CI operators is crucial for effective protection and resilience. The European Union Agency for Cybersecurity (ENISA) highlights a shortage of specialized trainings in areas such as Industrial

Table 6.1 Academic programs dealing with CIP

Country	Academic institutions	Programe	Level
USA	George Mason University Purdue University/ CERIAS University of Memphis	Public Policy with a focus on Leadership in CIP interdisciplinary covering risk management, incident response, and policy development multidisciplinary addressing technical needs of public utilities and emergency decision-makers	MSc/ 18m MSc MSc
UK	University of Bristol University of Surrey in collaboration with the University of Strathclyde Buckinghamshire New University Coventry University University of Derby	Cyber security (infrastructures security) Resilient digitalised critical infrastructure Critical infrastructure security Protective security and resilience Intelligence, security and disaster management	MSc/ 1y PhD/ 3.5y MSc/ 2y MSc/ 16m MSc/ 2y MSc/ 1y
Sweden	Luleå University of Technology KTH Royal Institute of Technology Swedish Defense University	Critical infrastructure protection Information security Cybersecurity Innovation, defense and security	MSc/ 6m MSc/ 2y MSc/ 2y MSc/ 2y
Netherland	University of Amsterdam/ Wittenborg University of Applied Sciences Leiden University Leiden University, TU Delft and The Hague University of Applied Sciences TU Delft, University of Twente and TU Eindhoven	Cybersecurity Crisis and security manangement Cybersecurity Cybersecurity	MSc/ 1y MSc/ 1y MSc/ 2y MSc/ 2y
Germany	University of Bonn University of Saarland Deggendorf University of Technology Augsburg University of Technology	Cybersecurity Cybersecurity Cybersecurity Industrial Safety and Security	MSc/ 2y MSc/ 2y MSc/ 2.5y MSc/ 18m

(continued)

Table 6.1 (continued)

Country	Academic institutions	Programe	Level
France	IMT Atlantique IMT Mines Alès	IT—Cybersecurity Disaster management and environmental impact	MSc/ 2y MSc/ 2y
Russia	ITMO University -The Faculty of Secure IT Moscow Institute of Physics and Technology	Several master's programs dealing with the CIP Cybersecurity	MSc MSc/ 2y

Control Systems (ICS) and Supervisory Control and Data Acquisition (SCADA) systems, which are essential for countering operational threats [17]. Furthermore, ENISA emphasizes the need for sector-specific training that considers the unique threats and dependencies of different CI sectors. This includes training in decision-making processes during data breaches or privacy incidents, which are often not covered in academic curricula.

USA like other countries with advanced CI protection systems often adopt integrated approaches that combine academic education with practical training. For example, the U.S. Cybersecurity and Infrastructure Security Agency (CISA) offers a wide array of free training programs to government and private sector partners, providing knowledge and skills needed to implement CI security and resilience activities [18].

6.3 Serbian Experience

6.3.1 Strategic and Legal Framework

The Republic of Serbia has recognized the importance of critical infrastructure protection (CIP) within its national security framework. The National Security Strategy of the Republic of Serbia [19] explicitly highlights the protection of critical infrastructure as a key security interest, particularly in the context of hybrid threats, cyber risks, and natural and man-made disasters. The Strategy defines critical infrastructure as assets and systems whose disruption would significantly affect national security, economy, public health, and safety.

Additionally, the Law on Disaster Risk Reduction and Emergency Management [20] provides a legal foundation for risk assessment and emergency planning, directly influencing the resilience of critical infrastructure. This law mandates the development of risk assessments and protection and rescue plans for all entities of special importance, including operators of critical infrastructure.

The Law on Critical Infrastructure [21]—This law regulates national and European critical infrastructure, identification and determination of critical infrastructure of the Republic of Serbia, protection of critical infrastructure, competence and responsibility of authorities and organizations in the field of critical infrastructure, information, reporting, provision of decision support, data protection, management and supervision in the field of critical infrastructure.

Sector-specific regulations in energy, transport, and ICT, which define obligations for operators regarding physical and cyber protection.

Collectively, these instruments frame the state's approach to CIP, with growing emphasis on intersectoral cooperation, risk-based planning, and integration of educational and training efforts.

6.3.2 Academic Education in CIP

Formal academic education related to critical infrastructure protection in Serbia is primarily offered within the broader fields of security studies, emergency management, and technical sciences. Although there is no standalone accredited study program fully dedicated to CIP, several faculties incorporate CIP-related content at undergraduate and postgraduate levels:

- University of Belgrade—Faculty of Security Studies:
  This faculty offers programs encompassing philosophical, sociological, legal, and technical aspects of security studies. It provides undergraduate to doctoral programs focusing on civil defense, environmental protection, and human security [22].
- University of Criminal Investigation and Police Studies:
  Provides specialized training in the field of national and public security, including courses related to the protection of strategic facilities and critical systems [23].
- University of Novi Sad—Faculty of Technical Sciences:
  The university offers a Master's program titled "Information Security in Critical Infrastructures," covering courses like critical infrastructure security, applied cryptography, and cyber incident analysis [24].
- University of Defense—Military Academy:
  Delivers modules in infrastructure defense, logistics resilience, and military aspects of infrastructure protection within its integrated military education system [25].
- Private higher education institutions (e.g., Union Nikola Tesla University, Megatrend University):
  Include security management programs with elective courses that partially address risk and infrastructure protection.

Postgraduate education in this domain is still evolving. Master's and doctoral research opportunities exist primarily through interdisciplinary programs combining security, ICT, and civil engineering components relevant to CIP. However, the lack of a dedicated, nationally standardized academic track in CIP remains a systemic gap.

6.3.3 *Licensed Training and Professional Development Programs*

In addition to academic pathways, several state institutions in Serbia provide licensed training programs relevant to CIP, particularly through the Ministry of the Interior and its subordinate agencies [26]:

- Sector for Emergency Management:
  Organizes certified training for public officials and infrastructure operators on emergency planning, industrial accident response, and risk management.
- Fire and Rescue Training Centers (Zemun and Niš):
  Offer practical education for disaster response personnel, including scenarios involving disruption of infrastructure systems.
- Centre for Basic Police Training – Sremska Kamenica:
  Provides education on infrastructure protection in the context of public safety, critical incident response, and anti-terrorism.
- Training programs for private security personnel, licensed under the Law on Private Security, include modules on technical and physical protection of critical facilities.

  It is important to emphasize that no private organization or institution in Serbia currently offers a state-licensed or accredited program specifically focused on CIP. While certain private training centers offer relevant content within broader security or risk management themes, these are not recognized as CIP-specific by national authorities.

6.4 Concluding Remarks

The educational landscape for CIP exhibits diversity in curricular content, pedagogical methods, and institutional focus. Most of the mentioned programs are one- or two-year master's degree programs, and a large number of them relate to Cybersecurity, i.e. the protection of KI from digital threats. Some programs are implemented in cooperation with several universities, and part of them are implemented by universities in cooperation with industry. USA institutions often integrate CIP into broader cybersecurity programs, emphasizing interdisciplinary approaches and practical applications. European institutions, while also interdisciplinary, tend to align CIP education with public policy and civil defense strategies. In contrast, institutions in Ukraine focus on integrating CIP into national security and emergency management frameworks, reflecting national security priorities.

Education and training in critical infrastructure protection are vital for developing a resilient and secure society. The comparative analysis reveals that while there is a shared understanding of the importance of CIP, educational approaches vary based on regional needs, institutional capabilities, and national security strategies. Continued collaboration and knowledge exchange among global institutions can enhance the effectiveness of CIP education and training programs.

However, academic education provides essential theoretical knowledge for CI protection, but it is insufficient on its own to prepare professionals for the complex, evolving threats to critical infrastructures. Supplementary, sector-specific training for CI operators is indispensable, ensuring that personnel are equipped with the practical skills necessary for effective risk identification, management, and crisis response. An integrated approach that combines academic education with continuous operational training is paramount for enhancing the resilience and security of critical infrastructures.

Finally, it must be borne in mind that the dynamic and unpredictable environment in which we live requires constant monitoring of new technologies, new risks and threats to AI and methods and ways of their protection, so that education and training must be constantly adapted, changed and improved so that the knowledge, skills and habits that are the outcomes of the teaching process are appropriate and adapted to the current moment and state of affairs. Lifelong learning in this field, as in a number of others in the modern world, is proving to be a necessity.

References

1. Kitagawa K, Preston J, Chadderton C (2017) Preparing for disaster: a comparative analysis of education for critical infrastructure collapse. J Risk Res 20(11):1450–1465. https://doi.org/10.1080/13669877.2016.1153501
2. Franchina L, Inzerilli G, Scatto E, Calabrese A, Lucariello A, Brutti G, Roscioli P (2021) Passive and active training approaches for critical infrastructure protection. Int J Disaster Risk Reduct 63:102461. https://doi.org/10.1016/j.ijdrr.2021.102461
3. Rajamäki J, Ruoslahti H (2018) Educational competences with regard to critical infrastructure protection. In: Jøsang A (ed.) Proceedings of the 17th European conference on cyber warfare and security (ECCWS 2018). Academic Conferences and Publishing International Limited, pp 415–423
4. George Mason University (2009) The CIP report, vol. 7, no. 12. https://cip.gmu.edu/wp-content/uploads/2013/06/83_The-CIP-Report-June-2009_EducationTraining.pdf
5. Chowdhury N, Gkioulos V (2021) Cybersecurity training for critical infrastructure protection: a literature review. Comput Sci Rev 40:100361. https://doi.org/10.1016/j.cosrev.2021.100361
6. Purdue University—Center for Education and Research in Information Assurance and Security (CERIAS) (nd). https://www.purdue.edu/campaigns/get-connected/
7. University of Memphis—Center for Information Assurance (CfIA) (nd). https://www.memphis.edu/cfia/

8. Newbold Jr KF, Delp BT (2011) Critical infrastructure protection program: a case study on higher education collaboration in homeland security. Homel Secur Rev 5(3)
9. University of Bonn (nd) Cyber security (M.Sc.). https://www.uni-bonn.de/en/studying/degree-programs/degree-programs-a-z/cyber-security-msc
10. Saarland University (nd). Cybersecurity (M.Sc.). https://www.uni-saarland.de/en/study/programmes/master/cybersecurity.html
11. Augsburg University of Technology (nd) Industrial safety and security (M.Sc.). https://www.tha.de/en/Electrical-Engineering/Industrial-Safety-and-Security-MSc.html
12. Deggendorf University of Technology (nd). Cyber security (M.Sc.). https://www.th-deg.de/cy-b-en
13. IMT Atlantique (nd). IT and cybersecurity (M.Sc.). https://www.imt-atlantique.fr/en/study/masters/msc/it-cybersecurity
14. IMT Mines Alès (nd). Master damage. https://www.imt-mines-ales.fr/nos-formations/master-damage
15. ITMO University (nd). The Faculty of Secure Information Technologies. https://abit.itmo.ru/en
16. The Moscow Institute of Physics and Technology (MIPT) (nd). Cyber security (M.Sc.). https://www.mastersportal.com/studies/275592/cyber-security.html
17. European Union Agency for Cybersecurity (ENISA) (nd). ENISA publishes training course material on network forensics for cybersecurity specialists. https://www.enisa.europa.eu/news/enisa-news/enisa-publishes-training-course-material-on-network-forensics-for-cybersecurity-specialists?utm_source#contentList
18. US Cybersecurity and Infrastructure Security Agency (CISA) (nd). https://www.cisa.gov/
19. Republic of Serbia (2020) The national security strategy of the Republic of Serbia. https://www.mod.gov.rs/multimedia/dodaci/prilog2_strategijanacionalnebezbednostirs_eng_1731678276.pdf
20. Republic of Serbia (2018) The law on disaster risk reduction and emergency management. https://disasterlaw.ifrc.org/media/3518
21. Republic of Serbia (nd) The law on critical infrastructure of Serbia and Europe. https://www.paragraf.rs/propisi/zakon-o-kriticnoj-infrastrukturi.html
22. University of Belgrade—Faculty of Security Studies (nd). https://en.wikipedia.org/wiki/University_of_Belgrade_Faculty_of_Security_Studies
23. University of Criminal Investigation and Police Studies (nd). https://www.kpu.edu.rs/en/
24. University of Novi Sad—Faculty of Technical Sciences (nd). https://isses.etf.bg.ac.rs/education-programs/
25. University of Defence—Military Academy (nd). https://www.uo.mod.gov.rs/eng
26. Ministry of the Interior (nd). https://www.mup.gov.rs/wps/portal/en

Part II
National Forward

Chapter 7
Adapting to the New Normal: Hybrid Threats and Critical Infrastructure Resilience

Nana Tabagua

Abstract This paper explores the evolving landscape of hybrid threats, emphasizing their impact on democratic resilience and critical infrastructure, with a specific focus on Georgia. Hybrid threats, characterized by a blend of military and non-military tools such as disinformation, cyberattacks, economic coercion, and covert operations, exploit systemic vulnerabilities to undermine democratic systems without triggering conventional warfare. Drawing on the CORE model developed by the EU's Joint Research Centre and the Hybrid CoE, the study highlights how these threats target the fundamental pillars of trust in democratic governance. Through analysis of two strategic cases, the Anaklia Deep Sea Port and the Namakhvani Hydropower Plant, the paper demonstrates how hostile actors, particularly Russia, have used hybrid tools to derail Georgia's strategic infrastructure projects and compromise its Euro-Atlantic aspirations. These cases reveal how legal ambiguities, weak institutional capacity, and strategic communication failures can be leveraged to erode public trust, polarize society, and exert geopolitical pressure. The concept of "Weapons of Mass Disturbance" (WMDi) is revisited to underscore the vulnerability of infrastructure to hybrid interference. Ultimately, the paper argues that hybrid threats pose the gravest danger to democratizing states, where incomplete institutional development makes them acutely susceptible to strategic disruption and external manipulation.

Keywords Critical infrastructure · Hybrid threats · Democratic resilience

7.1 Hybrid Threats: Concept

The greatest victory is that which requires no battle—Sun Tzu

In recent years, hybrid threats have become a major focus in Europe's security discussions. While some stakeholders may view them as a new challenge, they are

N. Tabagua (✉)
Policy and Management Consulting Group (PMCG), Tbilisi, Georgia
e-mail: n.tabagua@pmcginternational.com

S. Kentera et al. (eds.), *Building Resilience in a Dynamic Critical Infrastructure Landscape*, NATO Science for Peace and Security Series C: Environmental Security,
https://doi.org/10.1007/978-3-032-04383-2_7

far from a recent development. Hybrid threat tactics are, in essence, modern evolutions of age-old methods grounded in manipulation, psychological influence, and unconventional forms of conflict.

According to North Atlantic Treaty Organization (NATO), hybrid threats combine military and non-military as well as covert and overt means, including disinformation, cyberattacks, economic pressure, deployment of irregular armed groups and use of regular forces. Hybrid strategies are designed to obscure the line between peace and conflict, using ambiguity to create uncertainty and mistrust among targeted populations. Their core objective is to weaken societal stability and undermine public confidence.

While definitions of hybrid threats vary and need to remain flexible to respond to their evolving nature, the concept aims to capture the mixture of coercive and subversive activity, conventional and unconventional methods, which can be used in coordinated manner by state or non-state actors to achieve specific objectives while remaining below the threshold of formally declared warfare [1].

In the best-case scenario for the hostile actor, the combination helps to reach its strategic goals without detection, resistance and response. This type of tactic can exploit the opportunities that a changing security environment provides, as well as create new vulnerabilities and increase leverage for future use.

7.1.1 Hybrid Threats: Old Tactics, New Battlefield

Contemporary security challenges are characterized by a highly complex landscape, shaped by the ongoing war in Ukraine, persistent armed conflicts in the Middle East, escalating cyberattacks, state-sponsored disinformation campaigns, the strategic manipulation of energy resources and migration flows, as well as the disruptive implications of emerging technologies. These multifaceted factors, suggest that the occurrence and impact of hybrid threats are expected to escalate.

However, many experts argue that, the increasing prevalence of hybrid threats over the past 10–15 years stems not only from the evolving nature of security challenges, but also from the shifting dynamics of peace. Peacetime itself has transformed.

We are at the precipice of a world that is completely connected. From smart cities to smart buildings, with the use of machine learning, and deep learning we are closer than ever to being completely converged. This is also our Achilles heel which may lead to a disaster that we cannot possibly imagine [2].

Just as some have described the coming era as involving the 'weaponisation of everything', Hybrid Threats broad overarching toolbox gives great possibility to hostile actor to deliberately combine and synchronize action, to harm democratic societies. Unlike other concepts of new forms of conflict/warfare, the concept of Hybrid Threats, however, is the only one that raises the issue of systemic vulnerabilities of democratic systems as particular targets [3].

7.1.2 *The Comprehensive Resilience Ecosystem (CORE)*

Hybrid threats primary objective is to restrict the ability of democracies to act freely, ultimately undermining their credibility in comparison to authoritarian regimes or securing strategic advantages over them and to incur damage below the threshold of overt aggression.

As a part of the Comprehensive Resilience Ecosystem (CORE) developed collaboratively by the European Commission's Joint Research Centre and the Helsinki-based CoE for Countering Hybrid Threats, the core model structure of a democratic society was developed. The model encompasses thirteen domains—Political, Public Administration, Legal, Intelligence, Diplomacy, Military Defence, Infrastructure, Economy, Space, Cyber, Information, Social, Culture—each of which is considered vulnerable to hybrid threats. Hybrid tools can exploit or even create vulnerabilities within one or more domains, or at their interfaces. The objective may be achieved either through the direct impact of the tool on a domain or via cascade effects. Activity in one domain may be aimed at affecting a completely different domain from the one where the activity was detected.

The ultimate goal across these domains is to safeguard the core of the ecosystem: trust in the democratic system, which is built on seven key foundations:

- Feeling of justice and equal treatment
- Civil rights and liberties
- Political responsibility and accountability
- Rule of law
- Stability
- Reliability/availability
- Foresight capabilities.

Three of these foundations—stability, reliability/availability, and foresight capabilities, are heavily dependent on the critical infrastructure domain, which can be considered a 'mega-domain' due to its inclusion of multiple sectors.

7.2 Critical Infrastructure Targets—Weapons of Mass Disturbance (WMDi)?

While the protection of critical infrastructures primarily focuses on cyber and physical dimensions, they are often inadequately prepared for unconventional attacks by perpetrators of hybrid threats. This vulnerability makes them 'soft targets', potentially leading to economic or societal disruptions as severe as those caused by cyber or physical threats.

That is why Modern Critical Infrastructure seemingly serves as an effective instrument in the hands of adversaries able and willing to use hybrid tools.

The concept of Weapons of Mass Disturbance (WMDi) was first introduced in the context of terrorism threats. However, as the nature of threats in democratic countries has shifted from kinetic attacks to the exploitation of cyber and hybrid means, the **Hybrid CoE** revisited the WMDi concept in 2019 [4]. In an insightful working paper, the author contextualized WMDi within the broader challenge of hostile interference targeting critical infrastructure—an assessment that reflects both the urgency and relevance of emerging security dynamics.

Weaponizing Infrastructures is attractive for hostile actors as it gives possibility to coerce, intimidate, and apply pressure on a target state, as demonstrated by Russian cyberattacks against Ukraine's electric grid in 2015 and 2016, and against a range of critical infrastructure in the days and hours before Russia's invasion of Ukraine in February 2022.

Adversaries may employ kinetic means to achieve disruptive effects without engaging in overt military hostilities. A pertinent example of such a hybrid threat occurred in 2015, when Russian naval vessels repeatedly interfered with the installation of the NordBalt undersea electricity cable connecting Sweden and Lithuania. This high-voltage direct current (HVDC) link was designed to enhance energy security and market integration between the Baltic and Nordic regions, thereby reducing their dependence on Russian energy supplies. During the cable-laying operations, Russian warships entered Lithuania's exclusive economic zone on multiple occasions, ordering the construction vessels to vacate the area under the pretext of conducting military exercises. These actions prompted formal diplomatic protests from both Lithuania and Sweden, citing violations of international maritime law, specifically the United Nations Convention on the Law of the Sea (UNCLOS). The incidents exemplify how state actors can leverage military assets to exert pressure and disrupt critical infrastructure projects under the guise of routine operations.

Any tool capable of creating or exploiting vulnerabilities within an infrastructure to achieve strategic objectives of hostile actor could be considered part of the hybrid threat toolbox. The following compilation aims to provide a comprehensive overview of hybrid tools. However, it remains inherently open-ended, reflecting the constantly evolving tactics, creativity, and adaptability of hybrid actors:

- Direct Disruption and Physical Operations

  Physical operations on infrastructure; Destroying critical infrastructure components; Compromising service quality; Disrupting supply chains.
- Economic and Financial Manipulation

  Creating or exploiting economic dependencies; Foreign Direct Investment (FDI); Leveraging economic difficulties; Raising operational costs; Boycotts.
- Cyber and Information Warfare

  Cyber operations; Cyber espionage; Manipulating or spreading disinformation; Psychological pressure through service disruptions; Coordinated influence campaigns; Creating social division and unrest.
- Regulatory and Strategic Influence

Creating and exploiting infrastructure dependency; Exploiting legal thresholds; Exploiting regulatory loopholes; Influencing regulatory frameworks.

- Espionage and Covert Influence

Industrial espionage; Restricting or controlling access to vital resources; Pressuring demand; Infiltrating key institutions; Exploiting political instability.

7.3 Georgia—Playground for Russian Hybrid Aggression

In order to illustrate the practical implications of hybrid threats on critical infrastructure, this paper focuses on the case of Georgia. Its strategic geographic position, ongoing pursuit of Euro-Atlantic integration, and persistent exposure to external malign influence make it a relevant case for understanding how hybrid threats target and exploit systemic vulnerabilities.

Georgia was, in fact, the first target of Russia's hybrid warfare in 2008 [5], and since then, it has remained a battlefield for Russian hybrid aggression, having experienced nearly every tool in Russia's hybrid toolkit—making it, in many ways, a real-life laboratory [6].

Since 2003, Georgia has pursued closer integration with Euro-Atlantic institutions, including the European Union and NATO. However, the process has been marked by significant political and institutional challenges. In recent years, concerns have grown over democratic backsliding, political polarization, and the increasing influence of external actors, particularly the Russian Federation [7]. These developments have raised questions about the trajectory of Georgia's Euro-Atlantic aspirations and the resilience of its democratic institutions.

Despite Georgia's extensive exposure to various forms of hybrid aggression, this chapter focuses on two illustrative cases in which strategic and economic development, as well as energy independence and security, appear to have been adversely affected through a range of hybrid threat mechanisms. While no publicly available evidence can definitively confirm attribution in either case, the analysis remains theoretically grounded. Given that one of the primary challenges in addressing hybrid threats lies in their detection and attribution, the absence of conclusive data does not preclude meaningful examination of the patterns and impacts associated with such activities.

7.3.1 Anaklia Deep-Sea Port—Lessons from Georgia

Georgia's strategic location provides significant advantages, offering access to the Black Sea and serving as a critical link in the Middle Corridor, which connects Europe with Central Asia and China through the South Caucasus. Its significance has increased notably since the onset of the Russo-Ukrainian war, as the Middle Corridor has emerged as a key alternative to the disrupted Northern Corridor via Russia.

That said, the Middle Corridor remains constrained by substantial infrastructural and operational bottlenecks, which undermine its efficiency and limit its overall capacity [8].

In 2010, the Georgian government announced plans to construct the Anaklia deep-sea port, the project which holds immense strategic significance for Georgia, both economically and politically. It has potential to become a major trading hub between Europe and Asia. And what is most important Anaklia, serving as a critical connecting role between Europe and Asia, bypassing Russia, presents an ideal opportunity for the country to strengthen its relations with Western partners.

The port was initially scheduled to become operational by 2015. However, as of 2025, construction has yet to commence. This raises important questions regarding the factors contributing to the delay.

The artificial obstacles, particularly legal, regulatory and financial obstacles [9] created for the initial consortium, Anaklia Development Consortium (ADC) a Western-led consortium, including Maersk and Conti Group led to the contract termination in 2020, not only squandering a critical opportunity for regional hub, but also undermining Georgia's investment climate.

In 2023, following a three-year period of inactivity, the Georgian government released a new partner selection statement, ultimately awarding the Anaklia Port development project to a Chinese-Singaporean consortium led by China Communications Construction Company Limited (CCCC), a subsidiary of the state-owned China Communications Construction Group (CCCG).

China Communications Construction Co. (CCCC) is one of the main actors of China's Belt and Road Initiative, however the company has a questionable reputation, sanctioned by WB and many countries [10].

Given the increasingly close alignment between Russia and China, marked by strategic cooperation across political, economic, and security domains, Russia stands to benefit from China's expanding global footprint, particularly in regions of geopolitical contention. In this context, the involvement of a Chinese state-owned enterprise in a strategic initiative in Georgia raises serious security concerns. Such engagement not only risks deepening China's influence and economic leverage within Georgia, but also threatens to complicate the country's Euro-Atlantic aspirations. Over time, this could undermine regional stability and have broader consequences for security in the Black Sea area.

The following figure presents an analysis of the Anaklia Deep Sea Port case using the Hybrid CoE's CORE model framework (Fig. 7.1).

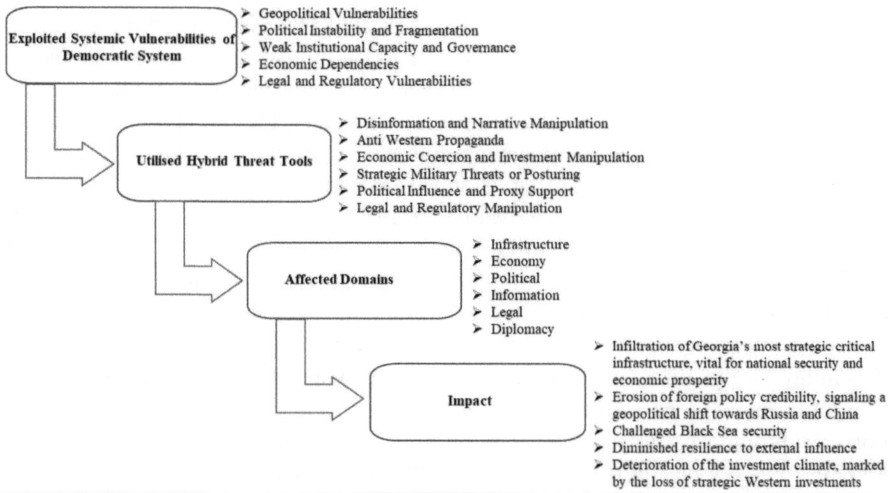

Fig. 7.1 Application of the CORE model to the Anaklia Deep Sea Port case

7.3.2 *Namakhvani Hydropower Plant—Lessons from Georgia*

Field experts have highlighted the Namakhvani HPP case as an example of Russia contributing to impeding a crucial energy infrastructure project in Georgia through propaganda and coercive economic tactics [11].

Notably, the Namakhvani project revealed numerous systemic shortcomings in Georgia's hydro resource management policies, alongside legal deficiencies and vague, unsubstantiated concessions. These issues triggered significant opposition from local communities, environmental activists, and international organizations. Stakeholders raised serious concerns about the project's potential environmental and social consequences. In particular, environmental groups conducted detailed analyses indicating that environmental considerations had been neglected throughout both the planning and implementation stages. They also highlighted violations of constitutional and property rights affecting local residents. Furthermore, documented infringements on rights related to access to environmental information, participation in decision-making processes, and the right to a safe living environment fueled widespread protests across the country. These tensions were further exacerbated by the Georgian government's ineffective strategic communication with the local population [5].

The case of the Namakhvani hydropower project illustrates how governance shortcomings created vulnerabilities, which effectively served as enablers of hybrid interference. In this instance, pro-Russian political actors and affiliated proxies effectively leveraged weaknesses in institutional transparency, legal accountability, and public communication to advance their agendas. Exploiting public concerns, these actors fostered distrust toward the national government, amplified fears regarding the foreign acquisition of Georgian land, and incited anti-Turkish sentiment, particularly targeting ENKA Renewables LLC, a Turkish industrial conglomerate and the project's principal investor. Ultimately, this coordinated disinformation and agitation campaign contributed to derailing a strategically important project (valued at USD 800 million) that could have significantly strengthened Georgia's energy independence and reduced its reliance on Russian energy sources.

The termination of the Namakhvani HPP project has had adverse implications beyond the immediate setback to Georgia's energy security and independence. It has also resulted in broader economic and reputational consequences, damaging Georgia's image as a reliable investment destination. This outcome has further eroded investor confidence and weakened the overall investment climate, potentially deterring future foreign direct investment (FDI), a critical driver of the country's sustained economic development. Additionally, in 2025, Georgia lost an international arbitration case brought by the investor, leading to a compensation payout of nearly USD 400 million—equivalent to approximately 2% of the country's GDP.

The following figure presents an analysis of the Namakhvani Hydropower Plant case using the Hybrid CoE's CORE model framework (Fig. 7.2).

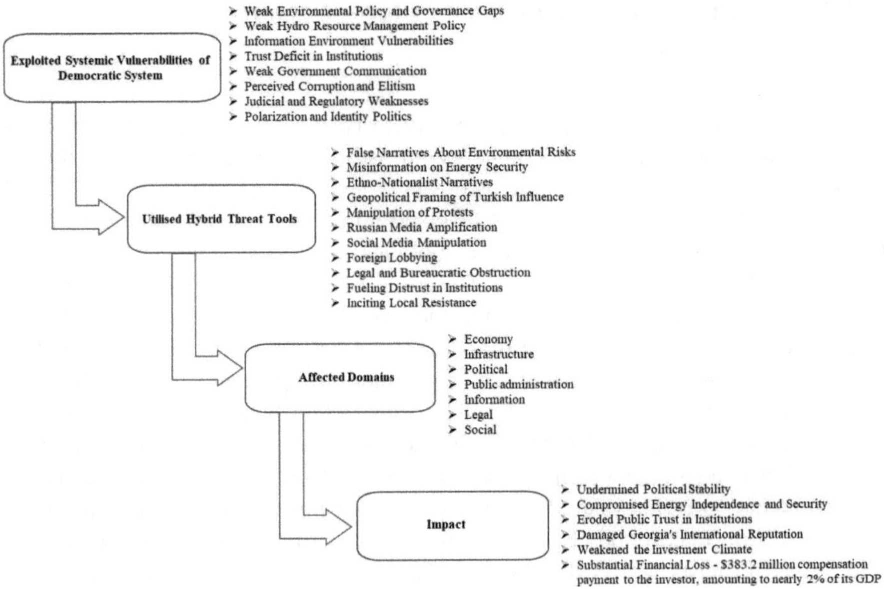

Fig. 7.2 Application of the CORE model to the Namakhvani Hydropower Plant case

7.4 Conclusion

To conclude, hybrid threats in the hands of strategically adept actors can yield devastating long-term consequences, even when their immediate effects remain obscured. By operating below the threshold of open conflict, these tactics systematically erode democratic institutions, disrupt critical infrastructure, and undermine public trust. Democratizing states, such as Georgia, are particularly susceptible. They often inherit the institutional fragilities of emerging democracies without the protective resilience afforded by long-standing governance traditions, making them prime targets for hybrid interference.

The case studies of the Anaklia Deep Sea Port and the Namakhvani Hydropower Plant clearly illustrate how hybrid tools, ranging from legal obstructions and disinformation to economic coercion, can derail strategic infrastructure projects and manipulate public sentiment. In such contexts, hybrid threats not only challenge national security but also threaten to shift geopolitical alignments.

In today's world, integrating hybrid threat mitigation into national security strategy is no longer optional, it is essential. Yet, in settings where political leadership is either beholden to foreign influence or prioritizes regime preservation over democratic consolidation, the risk is compounded. Without urgent, coordinated action to strengthen institutional resilience, strategic communication, and infrastructure protection, countries like Georgia may find their sovereignty and democratic aspirations increasingly at risk.

References

1. European Union (2016) Joint Framework on countering hybrid threats, a European Union response. Joint Communication to the European Parliament and the Council, EU Document JOIN
2. Bourgeix P (2019) Critical infrastructure security in a converged and interconnected world
3. Aho et al (2023) Hybrid threats: a comprehensive resilience ecosystem
4. Savolainen J (2019) Hybrid threats and vulnerabilities of modern critical infrastructure—weapons of mass disturbance (WMDi)?
5. Lazari A, Tabagua N (2024) Critical infrastructure security and resilience in Georgia
6. Seskuria N (2021) Russia's 'hybrid aggression' against Georgia
7. The 2025 worldwide threat assessment by the US Defense Intelligence Agency (2025)
8. World Bank (2023) Middle trade and transport corridor: policies and investments to triple freight volumes and halve travel time by 2030
9. Transparency International Georgia (2020) History of the investigations launched against the founders of TBC Bank and the Anaklia Port
10. The Commission on Security and Cooperation in Europe (U.S. Helsinki Commission) (2025) Bipartisan Helsinki Commission members express concern about Georgian dream awarding Anaklia Port development rights to Chinese Conglomerate. https://www.csce.gov/press-releases/bipartisan-helsinki-commission-members-express-concern-about-georgian-dream-awarding-anaklia-port-development-rights-to-chinese-conglomerate/
11. European Values Center for Security Policy (2021) Russian and Chinese influences in Georgia. https://mdfgeorgia.ge/uploads//Russian_and_Chinese_influences_in_Georgia-2021_update.pdf

Chapter 8
Review of Strategic Frameworks and Legal Regulations in the Field of CIP in Montenegro

Aleksandra Rakočević

Abstract The goal of the research is a studious and objective review of the legal aspect of critical infrastructure protection in Montenegro in order to modernize the normative basis and upgrade the protection system in accordance with the relevant EU acts in this area. The main research question is: To what extent is the legal framework for the protection of critical infrastructure in Montenegro compatible with European standards? We will get the answer to this question by using the comparative legal method, which enables the comparison of domestic legal regulations in this area with the legislation of the European Union and comparative legislation with the aim of improving national regulations and their application. The results of the research will indicate the necessity of innovating the existing national legal regulations and improving the normative framework for the protection of critical infrastructure. On the basis of a studious analysis of the legal basis, the adoption of new legal solutions in this area will follow. The conclusion is that only a modern legal framework harmonized with European standards and its consistent application can guarantee effective protection of critical infrastructure.

Keywords Montenegro · Critical infrastructure · Legal framework · European standards · Protection

8.1 Introduction

Numerous challenges faced by the world in the twenty-first century have highlighted the need to pay much more attention to the key capacities of every state, the endangerment of which inevitably leads to the paralysis of performing vital state functions. Among the most important potentials of a state undoubtedly is critical infrastructure, which ensures the regular functioning of systems and the servicing of citizens' needs.

A. Rakočević (✉)
Faculty of Law, University of Montenegro, Podgorica, Montenegro
e-mail: rakocevic.a@ucg.ac.me

S. Kentera et al. (eds.), *Building Resilience in a Dynamic Critical Infrastructure Landscape*, NATO Science for Peace and Security Series C: Environmental Security,
https://doi.org/10.1007/978-3-032-04383-2_8

Every attack on this public good leads to numerous material and combined consequences, which is why the state is obliged to create an optimal normative framework for the effective suppression and prevention of attacks on critical infrastructure in accordance with international standards.

As certain authors point out after the devastating terrorist acts in the USA, this kind of public good has become essential for the security of a state and is among the highest priority goals [1]. It is necessary for the uninterrupted execution of basic societal functions, with a focus on assets in a broader sense. It refers to the means that ensure the existence of multiple special segments from political to economic within the unified system, and any interruption in the performance of functions leads to irreparable losses. One of the most important segments of critical infrastructure is information and communication technologies, and compromising their functionality endangers the highest potentials of any state.

If we analyze the determination of critical infrastructure in EU acts, we will see that the content of the definitions is set more broadly and includes assets and systems of individual states without which the performance of essential social functions is not possible. This primarily refers to security, economy, health care, and the like. It rightly points to the need for mutual coordination so that within individual national systems, the implementation of protection measures does not have negative implications for other countries [2].

8.2 Legal Aspects of Critical Infrastructure Protection in Montenegro

National critical infrastructure in Montenegro in the previous period included state-owned enterprises in the fields of energy, transport, telecommunications, and similar sectors. Later, critical infrastructure came to encompass the network of state and public property protected by a system of social self-protection. Over a longer period, the critical infrastructure system included the defense sector, that is, large technical systems of particular importance for defense in the fields of telecommunications, information technology, transportation, water supply, electric power, and similar sectors.

It has long been an undisputed fact that the condition for the uninterrupted functioning of the state and efficient citizen services is the effective protection of critical infrastructure. This is why Montenegro has undertaken significant activities in planning and adopting regulations in the field of planning, building, and protecting the national critical infrastructure system. It should be noted that only partial examples in practice and theory refer to critical infrastructure as a unique and comprehensive system of national security, and efforts have been made to more comprehensively regulate this area of national security through legal solutions on critical infrastructure protection.

In December 2019, Montenegro adopted the Law on Determining and Protecting Critical Infrastructure [3], written in accordance with international agreements and EU standards. Article 2 of this document defines critical infrastructure in the following way: "systems, networks and facilities, or their parts located in the territory of Montenegro whose disruption of functioning, as well as interruption in the delivery of goods or services through these systems, networks, and facilities, may have serious consequences for national security, human life and health, property, and similar" [3]. The indicated determination of critical infrastructure essentially contains key elements of the EU 2008 definition, which, unlike some that define security threats for at least two EU member states, is included in the EU definition. If we analyze the content of this concept in other continents and global regions, there is a high degree of agreement that systems [4] such as water supply, energy, telecommunications, transport, and information technology are the most critical for the overall functioning of the state and society and must be considered and treated as critical infrastructure [5].

Chapter II, which deals with determining critical infrastructure, establishes criteria for determination as both sectoral and cross-sectoral criteria for identifying critical infrastructure. The criteria include an assessment of the consequences of disruptions in the operation or interruption of critical infrastructure in vital areas of public interest. In accordance with positive regulations, the Government of Montenegro prescribes sectoral criteria for determining critical infrastructure based on risk analysis and the characteristics of each sector in order to assess whether the impact of one sector on another could result in serious consequences. These areas include energy, transport, water supply, finance, healthcare, IT technologies, communications, environmental protection, and functioning of state authorities, defense, and the police as stipulated in Article 9 of the cited law [3]. The situation is similar in other countries, which almost without exception include energy, information and communication technologies, health, finance, transport, emergency services, water supply, defense, police et cetera under critical infrastructure [1].

Some authors emphasize the causal link between the telecommunications sector and other sectors, which is why this sector should be prioritized to increase the level of security [6].

Protection of critical infrastructure is regulated in Chapter III, which includes protection models, security plans, protection plans as security plans, as well as coordination tasks and handling of classified data. Protection models of critical infrastructure include the implementation of physical and technical protection similarly to the protection of persons and property, i.e., facilities and spaces. A special law regulates the methods of protecting critical information infrastructure used by the army, police, and the National Security Agency. According to Article 14 of the cited regulation, a list of systems, networks, facilities, and their parts of critical infrastructure significance is created based on risk analysis. Permanent security of critical infrastructure is a strategic and long-term goal of national security, and actions are taken depending on the level of risk and threats to the functioning of critical infrastructure. The coordinator performs protection tasks through monitoring the implementation of national and international regulations, monitoring the application and revision of the security plan, and mediating communication between operators and administrative bodies, etc. In case of critical infrastructure disruptions, a coordination team is formed in accordance with the Law on Protection and Rescue. Protection of critical infrastructure in the EU on the territory of Montenegro is regulated similarly.

The Law on Information Security of Montenegro [7], adopted by the Montenegrin Parliament at the end of last year, prescribes all necessary information security measures to achieve the highest level of security of network and information systems, including cybersecurity and management of this area, monitoring and supervision. The law is implemented by the Ministry of Defense, the Ministry of the Interior, and another body designated by the Government of Montenegro, which has access to public administration information systems and procedures with classified data. Its provisions do not apply to the Army of Montenegro, the National Security Agency, the Ministry of Internal Affairs, the Parliament of Montenegro, and the Central Bank of Montenegro. Article 11 prescribes two types of information security measures: the first relate to general protection of information systems, data handling rules, information system security measures, physical and technical protection; the second type of information security measures serve to protect against cyber threats, including physical protection, network and information system protection, and cyber risk management.

8.3 Institutional Aspects of Critical Infrastructure Protection in Montenegro

The institutional system for the protection of critical infrastructure in Montenegro is still at the very beginning of institutional development in this area. This area has traditionally been related to the defense of the state and included infrastructure of particular importance for defense, such as large technical systems, command centers, electronic systems, airports, ports, weaponry, military equipment, and the like. According to the Methodology for the selection of critical infrastructure adopted in 2014 by the Ministry for Information Society and Telecommunications of the Government of Montenegro, the following sectors of information infrastructure were identified: (i) information and communication technologies, (ii) finance, (iii) energy, (iv) healthcare, (v) agriculture and food safety, (vi) transport, (vii) water management, (viii) protection and rescue, (ix) state administration and public authorities, (x) education, (xi) industry. State authorities responsible for the protection of critical infrastructure include the Ministry of Defense, Ministry of Finance, Ministry of Economy, Ministry of Health, Ministry of Agriculture and Rural Development, Ministry of Defense, Ministry of Internal Affairs, Ministry of Justice, National Security Agency, and the Ministry of Transport and Maritime Affairs.

Within the Ministry of Public Administration, a Directorate for Information Security and Response to Computer Attacks has been established, tasked with early detection of cyber-attacks and threats as well as response to these dangers. Institutionally, this is the central body for coordination and protection from cyber-attacks. Within the Ministry of Internal Affairs – Police Directorate, a Group for Combating High-Tech Crime has been formed, tasked with preventing and suppressing criminal offenses against the security of computer data.

Montenegro, based on the Law on Information Security from 2024, envisaged the formation of an Agency for Cyber Security which is still not operational. The task of this agency is to protect the network and information systems of government bodies and other entities, particularly key and important bodies, except state bodies from cyber threats, serious cyber threats and incidents, as well as expert supervision over the application of information security measures with those bodies and entities, as stipulated in Article 6 of the cited law. This represents an important step forward in focusing the state on information infrastructure and preventive activity. Until now, the state has not had solutions regarding timely response to attacks, among other things, due to inadequate multisectoral and international cooperation.

Article 61 of the Law on Information Security foresees the formation of the Cyber Security Council as a body for monitoring the implementation of policies and strategies related to information security and proposing measures for improving the information security of the state.

8.4 Compatibility of Legal and Institutional Dimensions of Critical Infrastructure Protection in Montenegro with EU Standards

After the terrorist attacks on the European continent in the early twenty-first century, the European Council in 2004 requested the European Commission to develop a strategy for the protection of critical infrastructure. This resulted in the establishment of the European Critical Infrastructure Protection Program, which includes common protection principles, standards, definitions, a joint list of critical infrastructure sectors, priority protection areas, determined responsibilities of key stakeholders and a joint methodology [5]. Eleven critical infrastructure sectors were defined. The difference compared to Montenegrin legislation is that this national regulation does not recognize the private sector as a partner in critical infrastructure protection. Strong remnants of the past still dominate this area, and it remains entirely controlled by the state. It is worth noting that the role of the public sector in protecting critical infrastructure in the EU is extremely significant.

In terms of protecting critical information infrastructure, there are differences in the approaches of national and EU solutions. This is reflected in the lack of a strategic security approach when it comes to Montenegro, something the EU has been pushing since 2003. In Montenegro, there is no reliability or resilience of communication information systems. System institutions are not prepared for prevention, mitigation, or rapid response to incidents. A cyber security center has been planned, but is still not operational, and there is no public-private partnership to ensure the optimal functioning of the digital market. A major drawback in the area of technological standards in Montenegro is that the state does not have a defined system for the certification of cyber security at the market level. The state still lacks cybersecurity certificates and products and services, which would be a significant support for the realization of a single market in this area.

It can be stated that Montenegro still has not met the EU standards in terms of the organizational and institutional dimension of critical infrastructure protection. There are three requirements that operators and owners of critical infrastructure must apply in practice [8]. The first unmet condition relates to security measures, as a clearly defined protection plan by the operator who detects critical infrastructure resources does not exist. Considering that there is no such plan, it is neither regularly updated. The second requirement relates to risk assessment, and again, it is noted that Montenegro has not fully implemented this requirement, as it still lacks a comprehensive critical infrastructure risk assessment framework that would include all critical infrastructure segments. Without such a framework, the response cannot be effective in the event of a threat. The third condition relates to communication mechanisms. It refers to the way in which operators and critical infrastructure owners establish communication in case of incidents. This is also not fulfilled, as the establishment of adequate communication mechanisms has been very slow.

It is a fact that Montenegro has identified national critical infrastructure areas. It has also recorded systems and components of critical infrastructure. However, what is missing is an assessment of the interdependencies between established critical infrastructures. The country faces a serious obligation to develop emergency plans to timely protect national security.

8.5 Analysis of Attacks on the Information and Technology Network of Montenegro for the Period 2022–2025

With increased use of information technologies, Montenegro has become vulnerable in the area of information security. On August 20, 2022, the country was exposed to the largest cyber attack to date, which lasted several weeks. What is known is that the virus disabled information systems. Some 1000 computers were infected, and some government data ended up on the dark web, about 2.5 million files, including documents with sensitive personal data and confidential government materials. Contaminated stations were disconnected from the network and their hard drives were taken for forensic analysis. Officials stated that the attack was not random but a well-planned and financially prepared cyberattack. The National Security Agency announced right after the attack that it was orchestrated by foreign intelligence services. The agency warned Electric Power Company of Montenegro that it could be a target of a cyberattack, leading to a switch to manual system control at power plants. Days later, the Minister of Public Administration confirmed that the cyberattack was carried out by a well-known criminal group specialized in cybercrime—the Cuba ransomware group.

This type of ransomware limits access to the computer system or data by encrypting it and demanding ransom from the victim. Some types of ransomware block system access, while others encrypt files and demand ransom for the decryption key. Some do both. Victims are typically asked to pay a sum in exchange for file restoration. In such cases, data recovery is only possible by paying a financial amount or restoring from previously saved backups.

Specialists in cyber security information who published photos from the website of this hacker group made it clear that the virus could not have formed in just a month. According to certain sources, it has never been used in this format before. Remediation or network repair of certain servers is still ongoing, and the threat of the hacker attack has not passed. Some malicious software was found on workstations within the system, so the entire network had to be cleaned to prevent repeated attacks. A number of workstations are encrypted, which makes the computers unusable.

On the other hand, officials pointed out that there was a backup system, which limited the damage. However, the backup system was not built for such a large-scale incident. Some institutions used alternative emails such as Gmail or private accounts for communication, which is against legal provisions requiring official addresses on Montenegro's information infrastructure. The consequences of cyberattacks in Montenegro manifest as caused damage that may be financial, intangible—such as the exposure of confidential information—or combined.

In recent months, there have been several similar attacks in the region. After the attack on Albania's infrastructure, the country held Iran responsible and broke diplomatic ties with Tehran. Attacks were also recorded in North Macedonia, specifically on several ministries.

The database of the Statistical Office of Montenegro—Monstat—was hacked in mid-November 2024. Although this was persistently kept from the public for months, publicly available information shows that institutions were aware of the conclusions and risks. No current investigation results have been released. The known method used in the attack involves encrypting data and demanding ransom for its release. This form of criminal activity involves encrypting data in a way that makes it unreadable until the victim agrees to pay a ransom. This means hackers hold the data hostage, and even though the owners do not receive a decryption key without payment, they still cannot access it.

This criminal act legally justifies public concern due to the risk of unauthorized access or alteration of sensitive data, such as census records. In criminal and procedural terms, neither of these two hacker attacks has been resolved, meaning the perpetrators have not been identified or prosecuted.

8.6 Criminal Offenses Against the Security of Computer Data in Montenegrin Criminal Law

In Chapter XXVIII of the Criminal Code of Montenegro, criminal offenses in the field of computer crime are systematized.

This form of incrimination manifests itself in various forms of misuse of computer technology. Perpetrators of criminal activities jeopardize the flow and distribution of data, causing damage to vital values of society. Montenegrin legislation prescribes six criminal offenses against the security of computer data. The most common forms are: damage to computer data and programs, attacks on computer systems, the transmission and introduction of harmful computer programs, unauthorized data collection, unauthorized access to computer systems, and interference with the operation of computer programs.

One of the basic criminal offense is damage to computer data, which is prescribed in Article 349 of the Criminal Code of Montenegro [9]. The act consists of unauthorized deletion, alteration, damage, concealment, or any other action that renders computer data or programs unusable. The object of this offense is computer data or programs. The act of commission includes alternatively the unauthorized deletion, alteration, damage, concealment, etc. The consequence manifests itself in the destruction of computer data and programs. The offense is completed by taking actions that lead to the specified consequences. The perpetrator can be any person, and for culpability, intent is required that encompasses awareness of the essential elements of the act. If the same data is attacked with different actions, all those actions constitute a single criminal offense [10]. A more serious form of the offense qualifies with severe consequences, that is, the severity of the resulting damage, and for this, negligence is also sufficient besides intent. The law prescribes mandatory security measures of seizure of objects with which the offense was committed.

Another criminal offense is unauthorized access to a computer system (basic form of the offense: unauthorized access, deletion, alteration, damage, concealment, or any other action that renders computer data or systems unusable) [9]. The object of the criminal offense is computer data or the computer system. The act of commission is determined through the consequential norm as specified in the description of the incrimination. Insertion, as the primary form of the act, implies the placement of data or programs into a computer. Deletion means the destruction of computer data or programs. Damage to computer data or programs means making them unusable for their intended purpose. Altering a computer program or data means changing the content of the digital information it constitutes. Concealing data or programs means making them inaccessible to authorized individuals. The consequence of incrimination lies in the destruction or damage to the object of protection. The intention to disrupt the functioning of a computer system is the subjective element of the criminal offense. The perpetrator must have intent, and the offender can be any person [11]. A more serious form of this criminal offense exists if the disruption of the computer system affects data and programs important to state authorities, public services, business entities, etc.

The criminal offense of creating and introducing a computer virus includes two forms of execution. The first form involves creating a computer virus for a specific computer program or procedure. The second form involves the introduction of data that causes harmful effects using computer technology. It is known that computer viruses have the ability to reproduce and contaminate other data and programs. The consequences of their action are manifested in the alteration or destruction of data, loss of data, system disruptions, and so on. The criminal offense occurs when the introduction of the virus is intentional and causes significant disruptions. In this case, the criminal responsibility of the perpetrator arises, and the offense includes both the act of creation and the unauthorized insertion of the virus into the computer system.

When it comes to the criminal offense of computer fraud, it involves a special form of fraud where the act is carried out alternatively and includes influencing the results of electronic processing or data transfer, by entering incorrect data or failing to enter accurate data. The act includes the concealment or false representation of computer data and interference with the computer system with the intent of unlawfully obtaining material gain for oneself or another or causing material damage to another. The consequence of this criminal offense is the alteration of the results of electronic data processing. For this criminal offense to exist, the entered data must be false or there must be an omission in entering correct data. More serious forms of this offense occur when large material gain or damage is caused, while a lesser form exists when the offense is committed in a way that results in smaller material gain or damage or when an attempt is made to obtain such a gain. This criminal offense also requires intent, including the intent to obtain unlawful material gain or cause damage.

Another group of criminal offenses is listed under Article 353 of the Criminal Code of Montenegro. The basic form of this criminal offense includes unauthorized access to a computer system, which means full or partial use of the computer system without legal basis. The second form includes unauthorized access to a computer system by violating protection measures, that is, accessing a computer system that is of significance to state authorities, local self-government bodies, and institutions entrusted with public authority. In this case, the perpetrator acts contrary to the protection measures prescribed by law. The next form includes unauthorized interception of computer data, regardless of the method of their transmission, whether they are not public and are directed to the computer system, from it, or within it, including radio and electromagnetic emissions. The fourth form exists if the perpetrator uses or discloses data obtained through unauthorized access to a computer system, violating protection measures or obtaining them by intercepting computer data. A more serious form exists if the offense causes significant damage or if it is committed in a way that poses a serious threat to the functioning of the system. In every specific case, it is necessary to establish whether the perpetrator acted intentionally, which includes both intent and the use of data obtained by unauthorized access to a computer system.

The last criminal offense in this group is the abuse of devices or programs as stipulated in Article 354 of the Criminal Code of Montenegro. This criminal offense consists of the production, sale, procurement for use, import, distribution, or other methods of making available devices or computer programs designed or adapted for the commission of criminal acts of cybercrime, such as computer codes or similar data through which a system can be accessed in whole or in part for the execution of committed criminal offenses. Also incriminated is the possession of devices or programs with the intent of using them for the execution of criminal offenses in this area. Objects considered as devices or programs designed and adapted for the execution of cybercrime are also considered tools of the crime. The perpetrator must act with intent.

8.7 State and Trends of Cybercrime in Montenegro for the Period 2017–2025

Cybercrime refers to the misuse of information technology for unlawful purposes, which undermines the effective protection and functioning of IT technology. Today, during the commission of criminal offenses, computers and new technologies are significantly used, and computers often represent the object of illegal attacks. These are obvious negative implications of computer use (Table 8.1).

During the observed period, a total of 34 individuals were reported. The highest number of reported individuals was in 2020, with 9 persons (39.1%), followed by 2019 with 7 persons (30.4%). Due to the criminal offense of unauthorized access to computer systems, 5 persons were reported in 2020, and 2 in 2019. In 2017, 4 individuals were reported (Table 8.2).

In 2020, indictments were raised against 3 individuals. One individual was sentenced to prison. By the end of the year, 15 indictments remained unresolved [12]. These data show that digital transformation has led to an increase in cybercrime and its multiplication.

Table 8.1 Persons reported for committing criminal offenses against computer data security

Year	2017	2018	2019	2020	2021	2022	2023	2024
Number of reported persons	5	2	7	9	3	6	2	0
Total								34

Table 8.2 Indicted and convicted individuals for committing criminal offenses against computer data security in 2020

Indicted individuals	Convicted individuals	Type of sentence	Unresolved indictments
3	1	Prison	15

8.8 Methodology for Detecting Criminal Offenses of Computer Crime

Numerous criminal activities in this field include attacks on hardware, which implies destruction, damage, or theft of computer systems or their components. The same applies to software attacks, which in addition to the aforementioned forms include unauthorized alteration, publishing, or use of the software. Similarly incriminated are attacks on programs, data, system protection, and services.

The methodology for detecting these forms of high-tech crime is significantly different from resolving other types of criminal investigation. The situation is further complicated by the lack and shortage of specialized services that would facilitate faster and more effective problem-solving. High-tech crime leaves minimal traces and is mostly limited to electronic impulses in the form of electromagnetic signals. Criminal processing and subsequent investigation are made more difficult by the fact that perpetrators of these offenses use more advanced equipment than that possessed by official institutions, while most viruses are created based on new concepts.

After discovering that a criminal offense has been committed from this group, the police in cooperation with the state prosecutor's office apply certain criminalistic methods and techniques. In the first phase of criminal processing, the computer is disconnected from the system, and an integral analysis of all segments of the functioning of the computer system is conducted to determine the time, method, and means of execution with the help of experts in this field. After that, a criminalistic investigation follows, in which appropriate documentation and expert analysis are collected to provide material evidence. The suspects' or accused persons' computers and disks are confiscated for detailed forensic analysis. The copying of all data from digital media is carried out and stored securely for further use in investigations and criminal proceedings. The basis for proving these criminal offenses is a combination of material and personal evidence [13].

8.9 Conclusion

After obtaining the final benchmarks in the EU accession process, Montenegro aims to achieve very demanding standards in protecting critical infrastructure by strengthening its capacities and through intensive cooperation with European and NATO partners. The key goal is the implementation of the best European practices, which can contribute to the advancement of Montenegrin resources and competencies, in order to raise the level of security of critical infrastructure significantly. It is necessary for civil servants to be competitive and knowledgeable in the field of critical infrastructure security. Risks and threats to the security of critical infrastructure are increasingly pronounced, especially in the field of misuse of IT intelligence. It should be noted that cyberattacks on critical infrastructure were recorded in 2022, proving the need for an improved and strengthened system of critical infrastructure protection and timely detection of criminal offenses.

Montenegro has, through its historical development, gone through various stages of building critical infrastructure, which was used for the transportation of material goods by investing in road infrastructure, seaports, and airports. Today, Montenegro has a fairly well-developed critical infrastructure in the fields of information and communication technologies, energy, water supply, and transportation. Infrastructure systems are of great importance to the state and its citizens, and their complexity, interdependence, and vulnerability point to new security challenges in this area. However, Montenegro is still far from having sustainable, modern, and efficient infrastructure, which is why it is necessary to undertake more legal, organizational, and other measures to prevent the endangerment of critical infrastructure.

With the adoption of the Law on the Protection of Critical Infrastructure, Montenegro established a normative framework and defined the institutional resources, carriers, and responsibilities in protecting critical infrastructure. Numerous legal regulations have been adopted, including both laws and bylaws in this area. This mainly refers to the Law on Internal Affairs, the Law on Protection and Rescue, the Law on Municipal Police, the Law on the Protection of Persons and Property, and the Law on Weapons. Among strategic documents, the following should be highlighted: the National Security Strategy, the Cybersecurity Strategy, the Energy Development Strategy, the Transport Development Strategy, and others that significantly address the issue of critical infrastructure.

EU member states place great importance on the protection of critical infrastructure, which is confirmed by the EU Directive on the Protection of Critical Infrastructure from 2008. This document provides guidelines for Montenegro in terms of improving and protecting national critical infrastructure.

It is problematic that in Montenegro, the protection of critical infrastructure still lacks a proactive and preventive approach in terms of risk assessment. A multisectoral approach to this issue is also missing, which would include all relevant institutions in order to act preventively and avoid potential consequences in the event of critical infrastructure being endangered.

Modern threats increasingly come in forms that include cybercrime. Montenegro cannot be an exception and must participate in joint activities with other countries, especially as this often concerns transnational crime. The recent cyberattack, more than ever before, pointed to the need for adequate protection of critical computer infrastructure since the capacities of cyber defense proved inefficient. The hacker group in this case found weaknesses in the system through which they infiltrated and influenced the system to operate in a way that suited the hackers. This case showed that the capacities of cyber defense are still not sufficiently empowered and updated, which is particularly evident in crisis situations. Under conditions of asymmetric security threats to critical infrastructure, an integrated approach and intensive cooperation among all relevant entities are necessary. The greater the resilience of critical infrastructure to security threats, the better the state's readiness to respond to digital threats.

It is very important to emphasize the EU's commitment to solving problems in the field of cyber security as a condition for preserving and improving the digital connectivity of the old continent by allocating significant financial resources. The EU Cyber Security Strategy for the period 2020–2025 puts at the forefront the development of European resilience, independence, and operational capacities that would enable successful and efficient suppression of threats to information systems. Montenegro, in the process of joining the EU, has opened negotiation chapters 10 and 24 and continues to align its legislative framework with the EU acquis. In the last year's report on Montenegro, the EC emphasized the need for further strengthening of capacities in the field of cybercrime, but also the improvement of cooperation between key institutions at the national and international level.

In order to ensure an adequate, long-term and efficient information security system in the field of cyber security, it is necessary to consider extending the jurisdiction of state authorities to all criminal offenses where computers, computer networks, and data are used as the means of execution, or their products in digital or material form, which would enable more efficient identification of perpetrators and their criminal prosecution. It is most important to undertake preventive measures aimed at eliminating the sources of danger in this area, which implies the elimination of attacks or, if an attack has already occurred, minimizing the consequences. It is necessary to develop plans and strategies compatible with the highest standards in this field and to strengthen integrated security measures and approaches to the computer system, for which a thorough assessment of the computer system's vulnerabilities is required.

References

1. Škero M, Ateljević V (2015) Protection of critical infrastructure and basic elements of compliance with SE Directive 2008/114EC. Military Service, 3/2015
2. Council Directive 2008/114/EC, Official Journal of the European Union
3. Law on Determination and Protection of Critical Infrastructure, Official Gazette of Montenegro 7/2019
4. Murray T (2012) Critical infrastructure protection. Vulnerability Conundrum Telemat Inform 1:29
5. Lazari A (2014) European critical infrastructure protection (p 47). Springer
6. Rinaldi SM (2004) Modeling and simulating critical infrastructures and their interdependencies
7. Law on Information Security, Official Gazette of Montenegro 14/2010
8. Lindstrom M (2009) The European programme for critical infrastructure protection. In: Olson S (ed) Crisis management in the European Union. Cooperation in the face of emergencies (p 47). Springer
9. Criminal Code of Montenegro, "Official Gazette of RCG" No. 70/2003 and "Official Gazette of CG" No. 40/2008
10. Lazarević LJ, Vušković B, Vušković V (2004) Commentary on the Criminal Code of Montenegro, Cetinje
11. Rakočević V (2014) Criminal offenses with elements of organized crime—special research methods. Podgorica
12. Prosecutorial Council of Montenegro (2021) Report on the work of the Prosecutor's Council and the State Prosecutor's Office for 2020, Podgorica
13. Rakočević V (2012) Criminalistic methodology. Podgorica

Chapter 9
Critical Information Infrastructure Protection in Montenegro: Aligning National Regulations with EU Standards

Ivan Jeknić and Sanja Grbović

Abstract Critical Information Infrastructure (CII) represents the foundation of a digitized society and is a key element of national security in Montenegro, as a NATO member. In the modern context, CII is exposed to increasing cyber threats, necessitating adequate regulatory and technical mechanisms for its protection. The European Union has established a comprehensive framework for enhancing CII resilience through the Critical Entities Resilience (CER) Directive, the Network and Information Systems Directive (NIS 2), and the Artificial Intelligence Act (AI Act). The aim of this paper is to analyze the alignment of Montenegro's national regulatory framework with these European standards, identify regulatory gaps, and present proposals for improving national legislation.

Keywords Montenegro · European Union · Critical information infrastructure · CER directive · NIS 2 directive · AI Act

9.1 Introduction

Critical Information Infrastructure (hereinafter referred to as CII) forms the backbone of a digitized society and represents one of the key pillars of national security, economic development, and the stability of democratic institutions [1, 2]. CII encompasses systems, networks, and services whose disruption could have serious consequences for state security, the health and safety of citizens, and the uninterrupted operation of essential societal and economic functions [3]. Sectors considered part

I. Jeknić (✉)
Faculty of Political Sciences, University of Montenegro, Podgorica, Montenegro
e-mail: ivan.jeknic@gmail.com

S. Grbović
Faculty of Law, University of Montenegro, Podgorica, Montenegro
e-mail: sanjag@ucg.ac.me

S. Kentera et al. (eds.), *Building Resilience in a Dynamic Critical Infrastructure Landscape*, NATO Science for Peace and Security Series C: Environmental Security,
https://doi.org/10.1007/978-3-032-04383-2_9

of CII include energy, transportation, healthcare, water management, finance, digital communications, and public administration.

In the current landscape, marked by swift digital advancements and growing geopolitical uncertainties, Critical Information Infrastructure (CII) is exposed to a wide variety of threats. These include natural catastrophes, system malfunctions, and deliberate cyberattacks carried out by criminal organizations, hacktivists, or nation-states [4]. The increasing reliance of modern societies on digital technologies further complicates efforts to manage these risks, since disruptions in digital services can trigger ripple effects that impact multiple sectors at once.

Protecting CII necessitates a holistic strategy that combines legal, technical, and organizational actions, alongside strong collaboration between public and private sectors as well as between countries. Regulations in this domain need to evolve in step with technological progress, incorporating proactive systems for detecting threats, managing incidents, and enhancing resilience. Within this framework, aligning national laws with the European Union's legal standards is particularly important, especially for nations seeking EU membership or engaging in regional security partnerships.

The European Union has recognized the strategic importance of CII and developed a comprehensive regulatory framework establishing minimum standards for security and resilience in critical sectors. Key legal instruments in this area include the Directive on the Resilience of Critical Entities (CER Directive), the Network and Information Systems Directive (NIS 2 Directive), and the Artificial Intelligence Act (AI Act). Together, these instruments form the normative foundation for strengthening institutional capacities, technical preparedness, and strategic coordination in the protection of critical infrastructure.

The CER Directive [5] seeks to enhance the physical and operational robustness of organizations delivering vital services across key sectors such as energy, transportation, healthcare, water supply, finance, and public administration. It highlights the importance of conducting risk assessments, formulating national resilience plans, and recognizing critical entities. Furthermore, it mandates that Member States implement systems to prevent, prepare for, and respond to incidents that may jeopardize the security and ongoing operation of essential societal services.[1]

The NIS 2 Directive [6] updates and broadens the cybersecurity regulatory framework, succeeding the original NIS Directive from 2016 [6]. It sets more rigorous standards for cyber risk management, mandates the reporting of security incidents, and clarifies rules regarding supervision and enforcement. Additionally, NIS 2 expands its coverage to include more sectors such as public electronic communications, digital services, waste management infrastructure, food production, and critical technologies. This Directive plays a crucial role in establishing a consistent level of cyber resilience throughout the EU and strengthening cross-border collaboration among Member States via entities like the CSIRT network and EU-CyCLONe.

[1] The Directive introduces the Critical Entities Resilience Group (CERG), which is responsible for advising the European Commission, promoting cooperation among EU Member States, and supporting the exchange of relevant information, as outlined in Article 19(1).

The **AI Act** [7], while primarily focused on regulating the development and deployment of artificial intelligence systems, has direct implications for CII, especially where AI is used in high-risk applications within sectors such as energy, transportation, healthcare, and security systems [1]. The regulation introduces the concept of risk assessment for AI systems and sets out strict technical, ethical, and security requirements for those classified as "high-risk." This ensures that the use of AI does not introduce additional vulnerabilities into critical infrastructures but rather serves as a tool to enhance their resilience, efficiency, and predictive capabilities.

The following sections of this paper will focus on analyzing these three key legislative acts of the European Union, which collectively shape the normative framework for protecting CII. Through these complementary regulations, the EU establishes an integrated approach to CII protection, combining physical security, cyber resilience, and the secure application of advanced technologies. Harmonizing national legislation with these acts is not only essential for institutional and legal alignment, but also a prerequisite for effectively addressing modern security challenges in a digitally dependent society.

9.2 The Critical Entities Resilience Directive: A New Legal Framework for Infrastructure Protection in the European Union

The current geopolitical landscape is marked by heightened instability and escalating security challenges. The EU faces numerous threats, including Russia's ongoing aggression against Ukraine, an uptick in hybrid attacks and acts of sabotage, elevated risks of terrorism, and the growing impacts of climate change, such as severe weather events and water shortages [8]. In light of these developments, the Union must stay alert and continuously adapt its strategies.

On 16 January 2023, the European Parliament adopted Directive (EU) 2022/2557 on the resilience of critical entities (the "CER Directive"), thereby repealing Council Directive 2008/114/EC. The deadline for transposition of the Directive into national law was 17 October 2024. Additionally, Member States must identify and notify critical entities by 17 July 2026. The adoption of this Directive, alongside the updated Directive (EU) 2022/2555 (commonly referred to as the NIS 2 Directive), marks a new chapter in collaborative efforts at the EU level.

The CER Directive introduces legal obligations that are significantly broader in scope than its predecessor. While the 2008 Directive applied only to the energy and transport sectors, the CER Directive extends its application to eleven critical sectors, including banking, financial market infrastructure, digital infrastructure, energy, health, drinking water, wastewater, public administration, space, transport, and the production and distribution of food. This expansion is designed to rectify the lack of a harmonized approach among Member States, which previously led to divergent standards and an uneven regulatory landscape.

The CER Directive operates in conjunction with, but distinct from, Directive (EU) 2022/2555 ("NIS 2 Directive"). While both instruments share the objective of enhancing resilience, the CER Directive is focused on physical and operational risks, whereas NIS 2 primarily addresses cybersecurity threats. Importantly, the CER Directive incorporates an "all-hazards" methodology, requiring critical entities to prepare for a wide range of disruptive events, including natural disasters, terrorist acts, public health emergencies, and hybrid threats. The Directive thus represents a paradigmatic shift from reactive protection measures to proactive, preventive risk governance.

Under the CER Directive, critical entities, defined as those providing essential services the disruption of which would have significant societal or economic impacts, are subject to a suite of obligations. These include the conduct of comprehensive risk assessments (Art. 12) and the implementation of technical and organizational measures proportionate to identified risks (Art. 13). To support the consistent implementation of this Directive, Member States shall, where appropriate and without mandating or favoring any specific type of technology, promote the use of relevant European and international standards and technical specifications related to the security and resilience measures for critical entities. Member States are further required to ensure that entities conduct background checks on individuals occupying sensitive roles (Art. 13(2)(e)).

From a Member State perspective, the Directive imposes several substantive obligations. States must adopt a national strategy for the resilience of critical entities (Art. 6) and conduct sector-specific and cross-sectoral risk assessments (Art. 7). Furthermore, Member States, wherever appropriate, shall engage in mutual consultation concerning critical entities to ensure the consistent application of this Directive. (Art. 11).

Given these changes, organizations covered by the Directive should conduct a legal compliance review to assess their level of exposure and specific regulatory duties. This process involves mapping their services according to the Directive's sector definitions, carrying out internal risk evaluations, and adjusting current resilience strategies to meet the Directive's standards. Taking proactive steps is advisable, not only to reduce the risk of penalties but also to maintain operational stability and protect public confidence.

In conclusion, the CER Directive establishes a novel and legally binding structure for critical infrastructure protection within the EU. By aligning legal standards across Member States and broadening the definition of critical sectors, the Directive seeks to ensure a more resilient Union in the face of increasingly complex and transnational threats [9]. Effective implementation will demand coordinated efforts from both public authorities and private sector operators, and it is expected to become a benchmark for future legal advancements in infrastructure regulation and resilience management.

9.3 Implementation of the EU Council CER Directive in the Legislative Framework of Montenegro

Montenegro, as a NATO member state and a candidate for membership in the European Union, faces the imperative of establishing regulatory mechanisms to ensure the resilience and functionality of CII in accordance with European legal standards.

Montenegro's shift to a market economy and the privatization of critical infrastructure sectors have introduced various regulatory challenges. Despite changes in ownership, the state continues to be the main guarantor of infrastructure continuity and resilience by means of regulatory, supervisory, and strategic measures. In the field of information and communication security, progress has been made through the adoption of the Cybersecurity Strategy for the periods 2018–2021 and 2022–2026, as well as through the development of institutional capacities for protecting digital infrastructure [10, 11].

The Law on the Determination and Protection of Critical Infrastructure (2019) defines critical information infrastructure as systems whose disruption would have severe consequences for the basic functions of the state, activities of public interest and the life, health and safety of citizens. In addition, this Law introduces a systematic identification and classification of national and European critical infrastructure. However, this law still largely reflects the principles of Directive 2008/114/EC, which has now been repealed by the CER Directive. The CER Directive, as noted earlier, expands the sectoral scope (including healthcare, food, water supply, and digital infrastructure) and introduces obligations for all identified entities, including the development of resilience plans, risk assessments, and incident reporting [5].

To successfully implement this directive, Montenegro must undergo legal and institutional harmonization update existing laws, expand the sectoral scope, and provide technical support to operators. Cross-border cooperation and information exchange with EU member states are also essential in order to establish a coherent and functional resilience system [5].

Ultimately, complying with the CER Directive is both a formal obligation in Montenegro's EU accession journey and a crucial aspect of national security. Advancing reforms in this domain should be a strategic focus for Montenegro's public policy, risk management, and overall national resilience efforts.

9.4 The Network and Information Systems 2 Directive: The Cutting Edge of Cybersecurity Legislation in Europe

In the EU, CII protection is also established by the Directive on measures for a high common level of cybersecurity across the Union, more commonly known as NIS 2 Directive ("Network and Information Systems"). Network and Information Systems are defined as electronic communications networks, devices performing automatic data processing, as well as data contained in these networks and devices.

This Directive is the primary legislative act governing information security in the European Union. By its legal nature, it is not directly applicable in the legal system of a country (member state of the EU or non-member), and requires the country to adopt a specific regulatory act (or acts, in some cases) in order to directly enforce the provisions of the Directive within the legal system of the country ("transposition" into national legislation). These can be in the form of laws or secondary regulation (legal acts subordinate to laws, such as rules, regulations, orders, etc.). The adoption of the NIS 2 Directive was governed by the goals outlined in the EU Cybersecurity Strategy for 2020–2025 [2, 6].

The NIS 2 Directive regulates a wider scope of entities, not just CII operators (critical entities using ICT), and covers a variety of actors engaged in services and activities of public interest. These are divided into essential (which are considered of greater importance) and important entities (which are considered of relatively lesser importance). Essential and important entities as defined by the directive are entities engaged in essential services in sectors such as: energy and water supply, transport, banking, financial market infrastructure, public administration, digital infrastructure and services, waste water, space etc. The directive also requires member states to create a list of essential and important entities [12].

The NIS 2 directive provides for the establishing of a competent authority (one or more) for cybersecurity in each member state of the EU, including a cyber crisis management authority and a computer security incident response team (CSIRT). The provisions of the Directive allow for the creation of more than one competent authority at the national level, since there are many aspects of cybersecurity regulated by this directive: national cyber crisis management, computer security incident response, supervision and enforcement, different CI sectors, etc. [12]. The directive also establishes several institutions at the EU level, for the purposes of enhancing international cooperation, coordination, control and supervision. These are institutions such as the Cooperation Group, the European Union Agency for Cybersecurity (ENISA), the European cyber crisis liaison organization network (EU–CyCLONe), the Network of National CSIRTs, and a Committee of the European Commission [12].

The purpose of the directive is to establish protection against cyberattacks. At its core, this means applying legal and technical measures and systems to prevent, respond to and recover from cyberattacks. These measures are divides in two sections, cybersecurity risk management and reporting obligations. The Directive requires essential and important entities to engage in continuous and various cybersecurity risk management measures and their implementation. These include policies on risk analysis and information system security, handling of incidents business continuity (management of backups, disaster recovery, crisis management), supply chain security and risk assessment (direct suppliers or service providers), secure network and information systems (acquisition, development and maintenance, including handling vulnerability); effectiveness assess of these measures, issuing cybersecurity certificates (to entities complying with these legal requirements and standards), basic cyber hygiene practices and cybersecurity training, etc. The implementation acts containing technical requirements for these measures shall be adopted by the European Commission. The implementation of such measures must be in accordance with the relevant European and international security standards and technical specifications for network and information systems and ENISA guidelines [12]. Implementation acts, standards and other delegated acts are intended to be flexible and easily revised, in order to keep up with advancements in ICT, which is why they are drawn up and adopted by competent authorities of member states and EU authorities (such as the Commission and ENISA), and not elaborated in detail by EU regulations and directives.

Reporting obligations of cybersecurity incidents requires entities to inform competent authorities of ongoing or handled cyberattacks (threats and incidents). In case of a greater cybersecurity threat (cybersecurity crisis or incident involving two or more member states), competent authorities are obligated to report and share relevant information to other competent authorities of member states and competent EU institutions. Furthermore, the NIS 2 directive also provides for cooperation, mutual assistance and coordination between institutions for handling of large-scale incidents when necessary (for example, in case of a cybersecurity crisis or incident at the national or international level) [12].

Finally, the Directive requires member states to impose and enforce penalties (fines), as punitive measures against infringements of the provisions of the Directive and infringement against cybersecurity rules measures at the national level [12]. When essential and important entities use products with a digital component, the requirements, rights and obligations of users and competent authorities regarding the use and monitoring of these products are established by the Cyber Resilience Act [13]. Some of the competent authorities in this Regulation are the same as in the NIS 2 Directive [13].

9.5 Implementation of the NIS 2 Directive in the Legislative Framework of Montenegro

In Montenegro, the protection of critical information infrastructure is encompassed by the current Law on Information Security, adopted in 2024. This Law also regulates subjects outside of critical infrastructure sectors, but provides for measures and standards which must be applied by CII operators in their activities (operation and management of CII) [14]. Before exploring the provisions of this Law, it is necessary to briefly explore the history of the information security legal framework, in order to better review the current framework.

The first law fully dedicated to regulating this area in Montenegro was the Law on Information Security, adopted in 2010 (amended in 2016 with the adoption of the NIS 1 Directive). Before this legal act, these issues were regulated by subordinate regulations and indirectly governed by laws from other related areas [6, 15]. The significance of this Law lies in the creation of a legal framework for protection and security of information systems in public and private sectors. Based on this Law, two subordinate legal acts were adopted: one in 2010 and one in 2013 (a rulebook and a regulation), which elaborate on measures and standards for information security. However, two subordinate legal acts provided by this Law were never adopted: the regulation on the establishment and protection of critical information infrastructure and the regulation on the prevention and protection of information systems from computer security incidents (these matters were only regulated with the adoption of the new law in 2024) [15–17].

After a massive cybersecurity breach in 2022 on Montenegrin institutions, it was clear there were little to no measures in place to protect critical infrastructure and public service systems from cyberattack. More specifically, the procedure for responding to cyberattacks was insufficiently regulated, partly due to the failure to adopt the aforementioned subordinate regulations. Furthermore, a public opinion polling done in 2024 found that public trust in the security of electronic public services has declined by a significant margin after the 2022 incident, as citizens were unable to receive the requested information and electronic services on time [18, 19].

This has encouraged government authorities to create a new law, in order to better regulate these matters, and by the end of 2024, Montenegro had adopted a new Law on Information Security. Additionally, there was also a need to adopt a new law to align the regulation with the new NIS 2 Directive. Therefore, the new law was adopted for two express purposes: improving the protection of information systems of state bodies from cyberattacks, and for harmonizing the Montenegrin legal system with the NIS 2 Directive. Accordingly, the new law is fully aligned with the NIS 2 Directive and includes activities for identifying security risk levels (incidents), well defined security procedures and obligations, coordination measures, risk management, etc. In terms of text quality, the new Law on Information Security is a significant improvement over the previous act. In comparison to the old law, the new law has almost four times more provisions (the old law had 21 articles, the new law has 81). The new law also provides for the establishment of the Agency for Cybersecurity, a new body responsible for information security, along with the CIRT and Council for Information security [14, 19].

The CIRT is part of the Ministry of Public Administration, while the Agency for Cybersecurity and the Council for Information Security were founded to coordinate and monitor various aspects of information security, such as the implementation of the national strategy in this field, adoption of measures and procedures, etc. The CIRT is in charge of cybersecurity for state administration authorities, while the Agency is in charge of cybersecurity for other entities (subjects). All subjects are required to report cyberattacks to the CIRT or Agency, while the Ministry is responsible for the supervising the application of the Law. Larger cyberattacks (incidents and threats) are handled by the CIRT, the Agency, or both (depending on the scale of the incident) [14, 20, 21].

The Law on Information Security also establishes essential and important entities (called "key and important subjects" in the provisions of the Law) in accordance with the NIS 2 Directive. They are defined as subjects that use ICT for activities and services crucial to the functioning of the country and public interest activities, as well as to the life, health and security of citizens, including critical infrastructure operators [14]. These subjects are required by the Law to implement the full range of information security measures and standards in their operations. According to the Law, essential entities are subjects engaged in sectors of energy, water and food supply, traffic regulation, banking healthcare, ICT services, public administration etc. [14].

The Law provides for the adoption of a number of subordinate regulations for the implementation of this Law. The general deadline for adopting these regulations is 6 months from its entry into force (i.e., by June 5, 2025), with several specific deadlines for acts related to the Cybersecurity Agency. These subordinate acts have not yet been adopted, except for the Government's decision on the establishment of the Cybersecurity Agency. On the other hand, the Agency has not yet begun functioning, as it needs elected officials and staff to begin working (performing tasks such as reviewing cybersecurity protocols and measures of key and important entities, etc.), which have not yet been elected and acquired [14, 20, 22].

Information security in Montenegro is defined as the state of integrity, availability, confidentiality, and protection of data, and encompasses all activities necessary to protect information and network systems from cyber threats or incidents (cybersecurity of information and communication technologies—ICT). This concerns data collected, transmitted, stored, and processed through information and network systems, especially those in critical information infrastructure) [14].

Information security is achieved by applying appropriate standards and measures to establish and maintain data protection and protection from cyber threats or incidents (events and circumstances that compromise the information security of these systems and their data).

Data protection involves activities to remove and prevent damage from data loss, disclosure, or unauthorized modification. Specifically, data protection measures include rules for handling data, maintaining records of access to data, and the monitoring of data for security purposes [14]. Furthermore, information security also involves the protection of information and network systems, including their protection from cyberattacks.

Accordingly, measures to protect against cyber threats or incidents include the protection of data that is processed, transmitted, and stored in information and network systems, the protection of the integrity, confidentiality, and availability of data throughout the operation time span of these systems, as well as risk management in this area (adopting rules and procedures in case of cyberattack, etc.). Of course, physical protection of information and network systems is also necessary, which includes the protection of the space, facilities, and devices containing these systems. Information security measures are implemented in accordance with relevant standards for information security management ("MEST ISO/IEC 27001"). These standards were established long before the Law's adoption (the first version of the standard dates back to 2009, and the latest version is from 2023) [14, 16, 23]. Another requirement for the implementation of this Law is the establishment of institutions for a systemic response against cyberattacks, such as the Agency for Cybersecurity [24].

The current Cybersecurity Strategy of Montenegro provides for the formation of CII protection, as one of its operational goals. Montenegro does not possess adequate mechanisms and systems for detecting and responding to cyber threats, nor from recovering from these attacks. Additionally, CII protection was not adequately established in the past strategic period (prior to 2022). It is also noted that there is a global lack of experts in this area, especially for the public sector, and this problem is more severe in Montenegro. For these reasons, CII protection is a necessary goal in the strategic development of cybersecurity in Montenegro [11].

9.6 Artificial Intelligence Act: AI Systems in Critical Information Infrastructure

In the European Union, the use of artificial intelligence systems (hereinafter: AI) is governed by a dedicated regulation to protect health, safety, democracy, the environment, rule of law and fundamental rights and freedoms from the harmful effects of the use of AI systems—Regulation laying down harmonized rules on artificial intelligence (The Artificial Intelligence Act or "AI Act"). This regulation establishes rules and standards for the use of AI systems and their placement on the EU common market [7]. This includes a variety of rights and obligations for Union and Member State institutions and authorities, as well as users, producers and intermediaries of AI systems in the common market, particularly regarding high-risk AI systems [25]. Owners and producers of AI systems must ensure an adequate level of AI literacy among their employees. This includes expertise, experience, education and training for using artificial intelligence systems [7].

Furthermore, the regulation imposes legal requirements for using high-risk AI systems. These are AI systems used for processing biometric data, assessing job candidates, ranking students for educational admission, evaluating applications for temporary residence or visas (migrant rights), and more. This includes AI systems in critical infrastructure management and operation. Specifically, AI systems used as safety components in water, heating electricity supply systems, traffic regulation, as well as critical information (digital) infrastructure. Therefore, these AI systems are also considered high-risk [7].

Although the AI Act does not go into great detail on critical infrastructure, its provisions clearly impact public sector activities that rely on CII. AI systems deployed in critical infrastructure sectors (such as energy, transport, healthcare, water supply etc.) are considered high-risk systems and are thus subject to the most rigorous requirements set out in the Act, all of which are important for ensuring that AI systems used in critical infrastructure are both ethical and secure.

High-risk AI systems must meet technical and other compliance requirements with rules and standards (conformity) for production, use and placement on the market, as defined by the regulation (particularly for high-risk AI systems), to ensure they are resistant to cyberattacks (incidents) or failure. Conformity assessments are conducted by bodies registered with the competent authority of the Member State [7]. Specifically, the use of such systems must be accompanied by appropriate risk and data management measures (used by the AI system), cybersecurity, human supervision, etc. Additionally, high-risk AI systems must have prior technical documentation and built-in activity logs [7]. The regulation also allows for the processing of personal data of citizens, if certain conditions are met, in order to improve current AI systems. Among others, one of these conditions is that the AI system in question is used in critical infrastructure and networks [7].

Supervision in the area of AI technology is the jurisdiction of the European Commission and the AI Office, while in Member States, it is the responsibility of market surveillance authorities, as well as authorities for registering conformity assessment bodies. In these matters, the EU Commission is authorized to adopt various legal acts to further enforce and regulate in detail the provisions of the AI Act. The European Artificial Intelligence Board has also been established as a coordination body in this field. The regulation also prescribes penalties for violations of its provisions [7].

Other EU regulations also contain provisions regarding AI systems. For example, the Cyber Resilience Regulation requires that products with a digital element that qualify as high-risk AI systems must meet specific cybersecurity measures [7].

The development and use of AI presents specific challenges and ethical issues in activities related to critical infrastructure operations and management, such as protecting personal data, preventing discrimination, and ensuring transparency (for legal regulation of these issues in the EU AI regulation) [7]. In addition, AI technology evolves rapidly in unpredictable ways and can therefore act seemingly autonomously with unforeseeable outcomes [26–29]. Because if this, certain legal constraints are necessary for the protection of human rights and freedoms, such as legally mandated human supervision and built-in activity logs. Furthermore, without adequate supervision, AI system activities can easily exceed the legal authority of the user—for example, by accidentally collecting excessive personal data that is not legally authorized for processing. Careful legal regulation of AI systems is necessary during their integration and use in critical infrastructure operations and management.

There is also the question of how much AI regulation is needed? Overregulation in this area could slow down and hamper the development of AI in Europe, especially compared to the rapid advancement of AI in the United States and China [30, 31]. No matter the answer, some regulation is necessary to protect basic human rights and freedoms, as well as the public interest (one of which being CII). The direct application of the AI Act ensures strong regulation of these matters in the Member States, which can be adapted to new technological developments through legal acts of the European Commission and activities of supervisory and coordinating bodies.

9.7 Implementation of the AI Act in the Legislative Framework of Montenegro

As a non-EU country, Montenegro needs to adapt to EU requirements in the field of AI through its own legislation. There is no specific law on the use of AI systems in Montenegro, nor any other legal or strategic act regulating or governing the use of AI systems. On the other hand, the regulation of AI systems has been a topic some interest in Montenegro this year. Government representatives have announced ongoing efforts on the preparation and adoption of a Strategy for Artificial Intelligence on a national level (a guiding strategic document to foster a supportive environment for the use and development of AI technologies) [22, 24]. Furthermore, the local branch of the UNDP has published a report concerning the assessment for AI usage in Montenegro, in cooperation with the EU and Government of Montenegro [32]. However, until such a document is adopted (legal or strategic), there is no legal framework for the use of AI systems in critical information infrastructure operations and management.

In addition, a very high level of expertise and knowledge is required from employees who use, maintain, supervise, and improve AI systems for activities in the public interest (such as critical infrastructure operation and management). In fact, employees need skills and competencies that cannot be acquired solely through trainings and workshops, but require appropriate higher education and specialization in the use and function of AI systems [33]. This is especially important in critical infrastructure operations and management. Therefore, this is a matter which will require additional consideration when regulating AI usage in Montenegro.

9.8 Conclusion

In the legal framework of CII protection, there are three primary areas of improvement regarding the harmonization and implementation of EU law with Montenegro's legal system: the application of current law in the area of information security, the harmonization of current law in the area of critical infrastructure with the Critical Entities Resilience Directive, and the implementation of the AI Act. The first area of improvement requires the equal and continuous application of the current Law on Information Security, along with measures to achieve the goals and objectives outlined in strategic documents [2]. The next important step for CII security is the adoption of secondary legislation and other required acts, which are especially needed for the Law on Information Security, including the election of officials and employment of staff for the Agency for Cybersecurity, so the Agency can begin functioning. Additionally, CII protection in Montenegro cannot be achieved in isolation, which is why another important aspect is the coordination and exchange of information between competent authorities at an international level. Furthermore, the current Cybersecurity Strategy in Montenegro outlines the protection of CII as one of its operational goals. It remains to be seen how well this goal will be achieved at the end of the strategic period (from 2022 to 2026).

The second area of improvement is the adoption of national regulation for implementing the CER Directive. This can be with a new legal act, or with amendments to the current law (Law on the Determination and Protection of Critical Infrastructure). Either way, the adoption of secondary legislation and other required acts is also needed for the full implementation of the law and protection of CI in Montenegro. Additionally, the current Law in this field (unchanged since its adoption in 2019) is outdated in comparison to EU law (namely, the CER Directive), and the measures it imposed turned out to be ineffective against the great cyberattack on Montenegrin institutions in 2022.

The third area of improvement is also of great concern, as the Montenegrin legal system has no dedicated regulation of the use of AI systems. While there have been preparations conducted in Montenegro for the legal regulation of AI usage, the rate of progress is insufficient compared to the development of AI technologies. The AILA report is the most recent significant step towards the full legal regulation of this field of law, and there have been announcements that a national strategy on artificial intelligence is being prepared by Montenegrin institutions that is expected to be adopted in the near future. On the other hand, there has not been any significant strategic or legal document adopted in this area in Montenegro, for now. This is a very large deficiency in the legal system of Montenegro and should be addressed urgently.

References

1. European Commission (2021) Proposal for a regulation of the European Parliament and of the Council laying down harmonised rules on artificial intelligence (Artificial Intelligence Act). COM(2021) 206 final
2. European Commission (2020) The EU's cybersecurity strategy for the digital decade: joint communication to the European Parliament and the Council. JOIN(2020) 18 final, Brussels. https://digital-strategy.ec.europa.eu/en/library/eus-cybersecurity-strategy-digital-decade-0
3. Markopoulou D, Papakonstantinou V (2021) The regulatory framework for the protection of critical infrastructures against cyberthreats: identifying shortcomings and addressing future challenges: the case of the health sector in particular. Comput Law Secur Rev 41:105502. https://doi.org/10.1016/j.clsr.2020.105502
4. Adeyeri A, Abroshan H (2024) Geopolitical ramifications of cybersecurity threats: state responses and international cooperations in the digital warfare era. Information 15(11):682. https://doi.org/10.3390/info15110682
5. European Parliament and Council (2022) Directive (EU) 2022/2557 on the resilience of critical entities (CER Directive). Off J Eur Union L322:1–36
6. European Parliament and Council (2016) Directive (EU) 2016/1148 concerning measures for a high common level of security of network and information systems across the Union (NIS Directive). Off J Eur Union L194:1–30
7. European Parliament and Council (2024) Regulation (EU) 2024/1242 laying down harmonised rules on artificial intelligence (Artificial Intelligence Act) and amending certain union legislative acts. Off J Eur Union L198:1–156

8. Genini D (2025) Countering hybrid threats: how NATO must adapt (again) after the war in Ukraine. New Perspect. https://doi.org/10.1177/2336825X251322719
9. Pursiainen C, Kytömaa E (2022) From European critical infrastructure protection to the resilience of European critical entities: what does it mean? Sustain Resilient Infrastruct 8(sup1):85–101. https://doi.org/10.1080/23789689.2022.2128562
10. Government of Montenegro (2018) Strategija sajberbezbjednosti Crne Gore 2018–2021 (Cybersecurity Strategy of Montenegro 2018–2021)
11. Government of Montenegro (2022) Strategija sajberbezbjednosti Crne Gore 2022–2026 (Cybersecurity Strategy of Montenegro 2022–2026). https://www.gov.me/dokumenta/85e2a9d0-0d3c-483a-9822-515d3b7798de
12. European Parliament and Council (2022) Directive (EU) 2022/2555 on measures for a high common level of cybersecurity across the Union (NIS 2 Directive). Off J Eur Union L322:1–70
13. European Parliament and Council (2024) Regulation (EU) 2024/2847 of the European Parliament and of the Council of 23 October 2024 on horizontal cybersecurity requirements for products with digital elements and amending regulations (EU) No 168/2013 and (EU) 2019/1020 and directive (EU) 2020/1828 (Cyber Resilience Act). Off J Eur Union L347:1
14. Parliament of Montenegro (2024) Zakon o informacionoj bezbjednosti (Law on Information Security). Off Gaz Monten 113:24
15. Parliament of Montenegro (2010) Zakon o informacionoj bezbjednosti (Law on Information Security). Off Gaz Monten 14/10:40/16
16. Government of Montenegro (2013) Pravilnik o standardima informacione bezbjednosti (Regulation on Information Security Standards). Off Gaz Monten 019/13 (2013)
17. Government of Montenegro (2010) Uredba o mjerama informacione bezbjednosti (Regulation on Information Security Measures). Off Gaz Monten 058/10, 08.10.2010; 055/15, 30.09.2015
18. United Nations Development Programme (2024) Izvještaj, digitalizacija javne uprave: Ocjena zadovoljstva postojećim i identifikacija potreba za novim e-uslugama, opšta populacija – građani (Report, digitalization of public administration: assessment of satisfaction with existing and identification of needs for new e-services, general population—citizens). Third polling cycle, May 2024. CEED Consulting LLC Podgorica. In association with the Ministry of Public Administration and the UNDP. https://www.undp.org/sites/g/files/zskgke326/files/2025-03/01_izvjestaj_e-services_gradani_final_1707.docx.pdf
19. Ministry of Public Administration (2023) Izvještaj o sprovedenoj analizi procjene uticaja propisa: Zakon o informacionoj bezbjednosti (Report on the conducted regulatory impact assessment: Law on Information Security). https://wapi.gov.me/download/91065b50-29e7-4917-8ae0-6812d6444601?version=1.0
20. Government of Montenegro (2024) Odluka o osnivanju Agencije za sajberbezbjednosti (Decision on the establishment of the Cybersecurity Agency). Off Gaz Monten, 125
21. Government of Montenegro (2017) Odluka o osnivanju Savjeta za informacionu bezbjednost (Decision on the establishment of the Council for Information Security). Off Gaz Monten 48:17
22. Portal Vijesti (Portal News) (2025) Dukaj: Ministarstvo javne uprave ispunilo sve dosadašnje obaveze iz Reformske agende (Ministry of Public Administration fulfils all current obligations from the Reform Agenda). News article. https://www.vijesti.me/vijesti/drustvo/759673/dukaj-ministarstvo-javne-uprave-ispunilo-sve-dosadasnje-obaveze-iz-reformske-agende
23. Institute for Standardization of Montenegro (2023) Standardi MEST ISO/IEC 27001:2023, bezbjednost informacija, sajber bezbjednost i zaštita privatnosti (Standards MEST ISO/IEC 27001:2023, information security, cybersecurity and privacy protection). https://isme.me/sr_ME/project/show/isme:proj:101427
24. Ministry of Public Administration (2025) Hitno moramo nastaviti sa implementacijom Zakona o informacionoj bezbjednosti (Urgently continue with implementation of the Law on Information Security). Press release. Government of Montenegro. https://www.gov.me/clanak/hitno-moramo-nastaviti-sa-implementacijom-zakona-o-informacionoj-bezbjednosti
25. National Interoperability Framework Observatory (NIFO) (2024) EU, Digital public administration factsheet 2024. https://interoperable-europe.ec.europa.eu/collection/nifo-national-interoperability-framework-observatory/digital-public-administration-factsheets-2024

26. Milovanović D (2023) Upotreba veštačke inteligencije u javnom interesu – mogućnosti, izazovi i odgovornost (Use of artificial intelligence in the public interest—possibilities, challenges and responsibility). In: Međunarodni pravni odnosi i pravda: Zbornik radova 36. susreta Kopaoničke škole prirodnog prava - Slobodan Perović, tome 2, vol 2. International scientific conference, Kopaonik, 13–17 December 2023, pp 5–30. https://plus.cobiss.net/cobiss/sr/sr/bib/135962121
27. Toll D, Lindgren I, Melin U, Madsen C (2020) Values, benefits, considerations and risks of AI in government: a study of AI policy documents in Sweden. JeDEM EJournal EDemocracy Open GovMent 12(1):40–60. https://jedem.org/index.php/jedem/article/view/593
28. Bleher H, Braun M (2023) Reflections on putting AI ethics into practice: how these AI ethics approaches conceptualize theory and practice. Sci Eng Ethics 29(4):21. https://doi.org/10.1007/s11948-023-00443-3
29. Bostrom N (2018) Superinteligencija: Putevi, opasnosti i strategije. Akademska knjiga, Beograd
30. Sharma R (2024) AI regulation. In: AI and the boardroom. Apress, Berkeley, CA. https://doi.org/10.1007/979-8-8688-0796-1_3
31. Kashefi P, Kashefi Y, Ghafouri Mirsaraei AH (2024) Shaping the future of AI: balancing innovation and ethics in global regulation. Uniform Law Rev 29(3):524–548. https://doi.org/10.1093/ulr/unae040
32. United Nations Development Programme (2024) Artificial intelligence landscape assessment (AILA), shaping AI to be an empowering force for people and planet. Montenegro. https://www.undp.org/montenegro/publications/artificial-intelligence-landscape-assessment-aila
33. Pantiru MC (2019) Competencies necessary for eGovernment. National Agency of Civil Servants, Sofia, Romania. https://www.eupan.eu/wp-content/uploads/2020/02/2019-final-REPORT-Competencies-necessary-for-eGov-PRES-RO-1.pdf

Chapter 10
Challenges and Practices for Critical Infrastructure Protection with a Focus on the Republic of North Macedonia

Oliver Bakreski and Leta M. Bardjieva

Abstract Critical infrastructure protection is an essential element in the national security system and counterpoises a prerequisite for protecting regional and international security. Given the rapid technological development, the transboundary nature of threats to national security and, consequently, to critical infrastructure systems, as well as the cascading effect of the consequences of potential attacks on these systems, the imperative to build resilience is imposed. The focus of this paper is directed towards identifying the need for cooperation and coordination of the elements of the security system in the protection of critical infrastructure, in accordance with the regulations that determine its protection within the framework of regulatory predispositions with reference to the Republic of North Macedonia. The methodology for preparing this paper consists of applying a qualitative analysis of the legal framework and strategic determinations defined in specific goals for the protection of critical infrastructure, while the quantitative segment of the research for the preparation of this paper involves the interpretation of data obtained from a survey of the attitudes of a representative sample of respondents, members of private security in the Republic of North Macedonia, on the situation in the domain of the protection of critical infrastructure and the directions for its improvement. The research results elaborated in this paper highlight the real need to strengthen the available capacities in order to achieve an optimal level of security of critical infrastructure systems by completing the process of creating a legal framework for the protection of critical infrastructure and establishing guidelines for action towards achieving integration.

Keywords Legal framework · System · National security · Protection · Critical infrastructure

---

O. Bakreski (✉)
University Ss. Cyril and Methodius, Faculty of Philosophy, Institute for Security, Defense and Peace, Skopje, Macedonia
e-mail: oliverbakreski@yahoo.com

L. M. Bardjieva
European University, Faculty of Detectives and Criminology, Skopje, Macedonia
e-mail: bardjieva@gmail.com

S. Kentera et al. (eds.), *Building Resilience in a Dynamic Critical Infrastructure Landscape*, NATO Science for Peace and Security Series C: Environmental Security,
https://doi.org/10.1007/978-3-032-04383-2_10

10.1 Introduction

The quality of life in contemporary societies is directly affected by the ability of infrastructure networks to subsist with threats to their steadily functioning. This ability implies the capability to absorb the impact of potential catastrophic events and, as soon as possible, restore the conditions of normal functioning of critical infrastructure systems and, optimally, enhance resiliencePractices: A Comprehensive Guide to Learn Critical Infrastructure Security. (2024). (n.p.): Cybellium Ltd.".

Given the evolution and integration of the physical, cyber and social elements of critical infrastructure systems, there is an evident tendency towards regulation with the intention of building resilience of critical infrastructure systems. Hence, the definition, use and protection of critical infrastructure is a combination of processes [1].

The term critical infrastructure includes all objects, systems, networks and functions—physical or virtual, that are vital to the functioning of societies, and whose incapacitation or destruction would have a destructive impact on their survival [2].

In a contemporary context, the use of the phrase critical infrastructure protection was noted in the mid-twentieth century, during the Cold War and the Cuban Missile Crisis of 1962, with communications being identified as the first sector to be considered critical [3].

At the very beginning of the twenty-first century, this term was established in strategic defense and security terminology as a consequence of the 9/11 attacks in New York in 2001, with the White House preparing a National Strategy for the Physical Protection of Critical Infrastructures and Key Assets in 2003 [4].

The Cybersecurity and Infrastructure Security Agency (CISA) sector division identifies 16 sectors whose assets, systems, and networks, whether physical or virtual, are considered so vital that their disruption or destruction would have a debilitating effect on national security, economic security, public health, and public safety. These sectors include the chemical sector, commercial facilities, communications, dams, water and sanitation systems, critical manufacturing, defense industry facilities, emergency services, energy, financial services, food and agriculture, government facilities, healthcare and public health, information technology, nuclear reactors, and transportation systems [5].

As different businesses and industries become increasingly interdependent, the potential for a cascading effect of security breaches across multiple critical infrastructure sectors increases [6].

Consequently, regardless of the sector in which a particular segment of critical infrastructure is located in the geographical, legal and social context, the state can no longer be the sole entity that can independently provide protection in its entirety and relies heavily on the private sector for this purpose [7].

According to the International Security League, the private sector owns and protects approximately 85% of the infrastructure, while public security institutions have access to intelligence on threats to infrastructure [8].

According to Article 2 of Directive 2008/114/EC of 8 December 2008 on the identification and designation of European critical infrastructures and the assessment of the need to improve their protection, critical infrastructure protection is defined as all activities aimed at ensuring the functionality, continuity and integrity of critical infrastructures in order to prevent, mitigate and neutralize a threat, risk or vulnerability. Ever since, the process of identifying critical infrastructure systems at the level of nation states, as well as at the EU level, began, in parallel with the determination of instruments and mechanisms for their protection [9].

In line with INTERPOL's imperatives for critical infrastructure protection, intelligence sharing, capacity development and resilience building are essential elements in the integration of all relevant actors in preparing for, preventing and responding to threats and challenges that make the critical infrastructure system vulnerable [10].

In relation to the unification of coordination and integration efforts in critical infrastructure protection, there is a general consensus in the Ministry of Interior, the Ministry of Defense and the Chamber for private security in the Republic of North Macedonia, that according to the analysis of best practices and case studies, critical infrastructure protection counterpoises one of the most favorable areas for cooperation between public and private sector institutions, due to its span across multiple business sectors in efforts to strengthen national security, since attacks on critical infrastructure pose growing risks to societies [11].

In this regard, a cross-analysis of the legal framework and strategic provisions that determine the position of all relevant stakeholders in this process is being carried out, as well as the general discourse that prevails regarding the views on the level of cooperation and the directions for its deepening and strengthening in practice [12].

10.2 Critical Infrastructure Protection—Designating Strategic and Regulatory Positioning

Chronologically, the application of the term critical infrastructure in the normative-legal framework in the Republic of North Macedonia has begun to be regarded over the past decade, aligning with the country's efforts to harmonize its security and defense policies with the legislation of the European Union and NATO standards in order to regulate the institutional and coordination responsibilities for the implementation of tasks arising from the domain of critical infrastructure [13].

Thus, in 2016, the study published by the Chamber of the Republic of North Macedonia entitled Legal Framework for the Security of Critical Infrastructure—with a focus on the security of critical infrastructure is one of the primary publications that synthesizes the current situation in terms of defining, identifying and protecting critical infrastructure. This is where the indirect identification of critical infrastructure begins, primarily through the interpretation of the Decision of the Government of the Republic of North Macedonia on designating the legal entities that are obliged to have private security, especially defined in Chap. 4, which defines the private security for legal entities when it is in the interest of the security and defense of the Republic of North Macedonia [14].

Furthermore, in 2018, a significant milestone was the adoption of the National Cyber Security Strategy for 2018–2022, which explicitly addresses the protection of critical information infrastructure as a strategic goal [15].

The 2020 Defense Strategy, defines critical infrastructure vulnerability as one of the main threats to national security, for which, in relation to the identified security challenges, threats and risks, the defense system of the Republic of North Macedonia commits to building capacities and capabilities to respond to threats to critical infrastructure, as defined in the defense interests and defense objectives [16].

As part of the government's reform agenda in the process of accession and harmonization of national legal acts with the law of the European Union (EU) and further building capacities for the successful implementation of the National Program for the Adoption of the Acquis (NPAA), the Ministry of Defense (MoD), in cooperation with the Secretariat for European Affairs (SEA) and foreign and domestic expertise, in the last quarter of 2021 began preparations for regulating the obligations related to the regulation and protection of critical infrastructure [17].

Further affirmation of the need for institutionalization of critical infrastructure protection occurred in 2022, in the proposal for the law on critical infrastructure protection in the Republic of North Macedonia, where the definition provided that critical infrastructure is that whose interruption or disruption affects key societal lifelines, which has negative implications for people's health and lives, the environment, citizen safety, property, transportation, water and food supply, energy, communications. Thus, the solution is offered by which critical infrastructure is defined as assets, systems, facilities, networks and other elements on which society relies on in maintaining national security, economic vitality and public health and safety [18].

Furthermore, the proposed legal provision states that planning, organizing and implementing the protection of critical infrastructure and ensuring the maximum level of security is a key priority in tackling a wide range of threats and security challenges of various nature with various levels of intensity and different carriers that directly or indirectly endanger critical infrastructure in the Republic of North Macedonia. From the perspective of threats, this taxonomy inevitably includes the vulnerabilities of critical infrastructure systems as well as the challenges arising from the need for adaptation and resilience to the onslaught of natural disasters and climate change [19].

Also, the National Development Strategy 2024–2044 as a strategic and development document, identifies key challenges and state priorities, especially considering the aspiration of the Republic of North Macedonia to join the European Union (EU) by 2030 and the commitment to sustainable development embodied in the 2030 Agenda. The National Development Strategy basically includes the main objectives defined in the adopted strategic documents, but also provides guidance for future strategic documents and policies. In this strategic document, the hazard profile of the Republic of North Macedonia is characterized by the impact of a range of natural and man-made hazards, with floods, earthquakes, extreme weather events, landslides, droughts, epidemics and technical–technological incidents dominating. According to the strategic area: Safe, Secure and Resilient Society in the Strategy, the Republic of North Macedonia, like other countries, faces vulnerability to cyberattacks targeting key sectors such as infrastructure, government, education and business, which cause service disruptions and economic damage [20].

The Cybersecurity Strategy 2025–2028 is a strategic document of the Republic of North Macedonia, the aim of which is, through a range of activities and measures, to ensure a safe, resilient and trustworthy digital environment. Here, the security of information systems (cybersecurity) is defined as a key national priority for the protection of digital infrastructure and services used by citizens, state and public institutions and the economy. The basic principles on which the Strategy is based, among others, include coordination, cooperation and support at the national and international levels, security by design and based on risk assessment, through the application of a holistic approach. This strategy and the Action Plan (which is a component document of the Strategy) define the roles of the various stakeholders in cybersecurity, strengthened cooperation between the public and private sectors and the quality management of professional staff in this area at the national level. As an integral part of the National Cyber Security Framework for effectively addressing risks in cyberspace, the Republic of North Macedonia is committed to building cyber defense capacities according to the highest standards. The development of cyber defense capacities is focused on the Army and has strategic importance for the overall national defense of the country [21].

For these reasons, the Army is developing a military team for responding to computer security incidents and capabilities for deploying cyber operations. In this context, within the framework of the acceptance into the NATO Cyber Defense Center in Tallinn, Estonia, the Republic of North Macedonia will be able to use the Center's tools to strengthen its own resilience against hybrid attacks and strengthen the protection of critical infrastructure [22].

In the context of the security environment and the general discourse, critical infrastructure has been one of the top priorities of the NATO alliance since the Warsaw Summit in 2016 [23]. At the Warsaw Summit, the joint engagement between NATO and the European Union in the area of critical infrastructure protection, through the adoption of a joint declaration, identified key areas for expanding cooperation, including addressing hybrid and cyber threats, and supporting partners in building defense capacities.

Specifically, this means defining the key areas according to which the NATO alliance assesses resilience in each member country, which include: continuity of government and critical government services; resilient energy supplies; resilient food and water resources; the ability to effectively manage uncontrolled migration; the ability to deal with mass casualties, as well as resilient civil communication systems [24].

Consequently, at the Madrid Summit in 2022, the Heads of State and Government of NATO member countries adopted the Strategic Concept, which emphasizes that ensuring national and collective resilience is key to all essential tasks and is the basis of the commitment to protect the nations of the Alliance member countries, societies and shared values [25].

With the Commitment to Strengthen NATO Resilience, Allies commit to continuing to strengthen NATO's resilience as the first line of defence and to developing individual and collective capacity to counter any type of attack. Civilian preparedness counterpoises the central pillar of Allied resilience and a key enabler of the Alliance's collective defense [26]. Civilian preparedness is improved by achieving NATO's core requirements for national resilience, which focus on continuity of governance, continuity of essential services, security of critical civilian infrastructure, and support of military forces with civilian assets.

In this regard, within the framework of the Long-Term Plan for Defense Development and Confirmation of the Defined Goals, Capabilities and Capacities of the Army, 2023–2032, priorities are defined, including the implementation of the seven NATO basic requirements for resilience, through the establishment and support of national goals for resilience to threats and risks for the Republic of North Macedonia [9], where in addition to conventional and transnational threats, asymmetric, hybrid and cyber threats represent an increased security challenge [27].

From the perspective of the Republic of North Macedonia's commitment to full EU membership, the significance of critical infrastructure also stems from the fact that it is covered in a large number of chapters of the acquis, especially Chap. 3.24 Justice, Freedom and Security, 3.10 Information Society and Media and 3.31 Foreign Security and Defense Policy. Given these reasons, timely regulation and designation of the protection of critical infrastructure is of particular importance in the approximation process [28].

According to the National Program for the Adoption of the Acquis of the European Union (NPAA 2021-2025), within the framework of harmonization of the regulation regarding critical infrastructure, the adoption of a Law on the Protection of Critical Infrastructure is envisaged, as well as the determination of executive competence within the Government of the Republic of North Macedonia [29]. In this regard, the draft law on the protection of critical infrastructure was proposed as a top national priority at the initiative of the Ministry of Defense, in order to build an effective strategy for the protection of critical infrastructure in the Republic of North Macedonia [30]. Given that the Ministry of Defense has the required expertise within the framework of the seven basic criteria for resilience in the NATO context, and in accordance with the proposals noted in the Integrated Report on State Reorganization [31], which foresees the integration of several directorates and centers, it was determined that the coordination role in the process of drafting the Law on Critical Infrastructure would be taken over by the Ministry of Defense [32].

Therefore, within the framework of supporting the government's reform agenda in the process of accession and harmonization of national legal acts with the law of the European Union (EU) as well as identifying the need to improve the protection of critical infrastructure and the implementation of Directive 2022/2557/EU of the European Parliament and of the Council of 14 December 2022 on the resilience of critical infrastructure entities and further building capacities for the successful implementation of the National Program for the Adoption of the EU Acquis (NPAA), the

Ministry of Defense (MoD) in cooperation with the Secretariat for European Affairs (SEA) has begun preparations to regulate the obligations related to the regulation and protection of critical infrastructure [33].

As part of the commitments and as part of the obligation to align with the legislation of the European Union, in the part of Chapter 24—Freedom, Security and Justice, the candidate countries for full membership of the European Union should implement the alignment with Council Directive 2008/114/EC prior to their official accession to the Union. The Directive establishes the procedure for the identification and designation of European critical infrastructure, primarily to improve the prevention, preparedness and response of the European community to terrorist attacks involving critical infrastructures, where the primary and ultimate responsibility for the protection of European critical infrastructure lies within the Member States and the owners/operators of such infrastructures.

In this regard, the Joint Action Plan on Combating Terrorism in the Western Balkans between the European Commission and the authorities of the Republic of North Macedonia also aims to strengthen the protection of citizens and infrastructure. In particular, this commitment refers to the engagement of each Western Balkan partner to improve the protection of their public spaces in accordance with the European Action Plan in support of the protection of public spaces, as well as their critical infrastructure [10].

Consequentially, the identification and determination of the indicative list of sectors and the identification of critical infrastructure facilities is to a certain extent carried out through the interpretation of the Law on Private Security, where Article 44 on mandatory security stipulates that the Government of the Republic of North Macedonia designates the legal entities obliged to have private security [34].

This article is complementary to the Decision of the Government of the Republic of North Macedonia on designating the legal entities that are obliged to have private security, especially defined in Chap. 4, which defines the private security of legal entities when it is in the interest of security and defense, in handling objects and facilities of particular cultural and historical significance, as well as for legal entities on the territory of the Republic of North Macedonia registered to handling radioactive materials or other substances hazardous to humans and the environment [35].

Hence, the Strategy for the Development of Private Security in the Republic of Macedonia 2016–2020, in accordance with the Law on Private Security in the Republic of Macedonia from January 2017 [36], sets out the strategic goals that are planned to be realized and which relate to public–private partnerships and the protection of critical infrastructure Fig. 10.1.

From the perspective of fully completing the process of legal and regulatory designation of critical infrastructure as a segment of strategic management, the intention is to provide a legal basis for further development of institutional cooperation at the level of public administration institutions and the private sector, especially the private security sector as part of the national security and defense system [37]. Undoubtedly, the private sector in the field of security is part of the state's security system, and therefore, in terms where institutions are required to be rational and efficient in the use of financial and human resources, the question arises of how a system can be

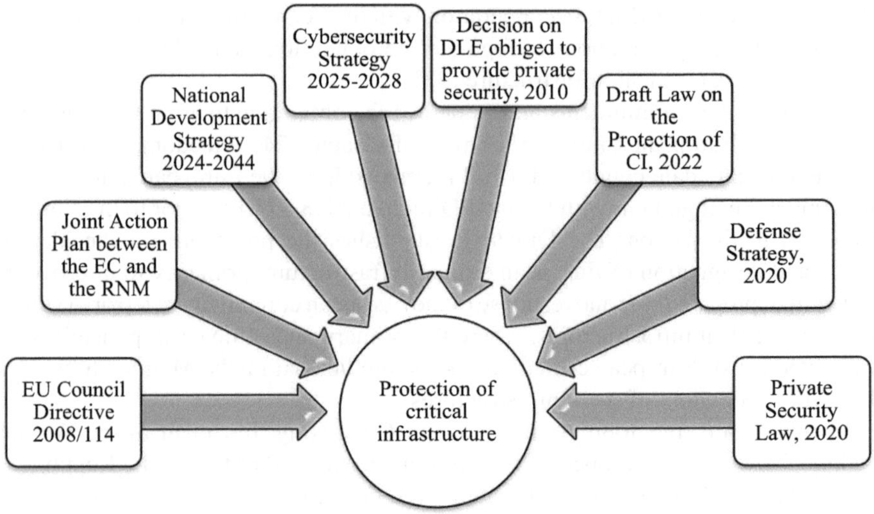

Fig. 10.1 Regulatory and legal framework for critical infrastructure protection in the Republic of North Macedonia

jointly built that will primarily serve the state, and citizens will be the ones who will benefit, as the European Commission pointed out in 2024 [38], regarding the need for joint operational cooperation between the public and private sectors. This operational cooperation, within the framework of specific projects and public–private partnerships, is also correlated with a higher level of institutional cooperation [39].

10.3 Surveying the General Discourse of Stances on an Operational Level

According to the results of the qualitative analysis and the review of the legal provisions and strategic determinations that regulate this segment, the emphasis is identified on one of the best practices in building resilience of critical infrastructure systems, especially from the aspect of inter-sectoral and inter-departmental cooperation, which is the exchange of data and information.

Hence, the quantitative part of the research for the purposes of this paper offered an introspection of the opinions of private security personnel in the domain of the contribution of the private security industry to the protection of critical infrastructure. A representative sample of 73 respondents was sent a Questionnaire with structured questions through whose answers, among other things, their views on the degree of completeness of the legal framework of the private security industry, the level of training and authorizations of private security personnel for the protection of critical infrastructure, as well as operational cooperation and exchange of information with state security institutions, through a process of operational coordination and a holistic approach in order to ensure the security and protection of critical infrastructure [40].

Thus, to the question formulated as: "Do you consider that private security can effectively manage the threats and risks to critical infrastructure?", where the offered answers ranged within these frames: (1) yes completely, (2) yes, partially, (3) sporadically (4) no or (5) not sure; the data depicts that 33 of the respondents believe that private security, with its capacities, authorizations and tasks, can fully deal with the threats and risks imposed to critical infrastructure, while 26 respondents, in their opinion, believe that members of private security can partially deal with the threats and risks facing critical infrastructure.

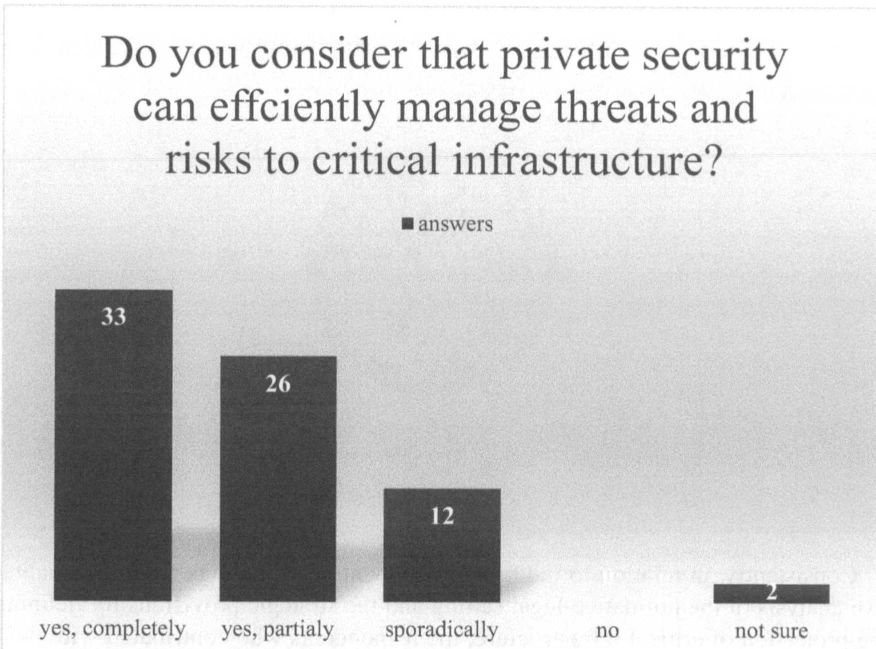

In the same context, in line with the tendency to integrate the engagement of all stakeholders in the security sector in responding to contemporary risks and threats to the security of critical infrastructure, the question was asked: "How significant is the cooperation and coordinated effort of all security entities (police, army, intelligence services, private security) in dealing with potential risks and threats?", whereas the following possible answers were offered: (a) very significant, (b) partially significant, (c) significant to a certain extent and (d) don't know.

The answers to this question emphasize the general position of private security personnel on the necessity of coordinated engagement of all security entities that are part of the overall security system of the state in dealing with potential risks and threats to public safety, where 85% believe that cooperation and coordinated efforts of all security entities (police, army, intelligence services, private security) in dealing with potential risks and threats is very important.

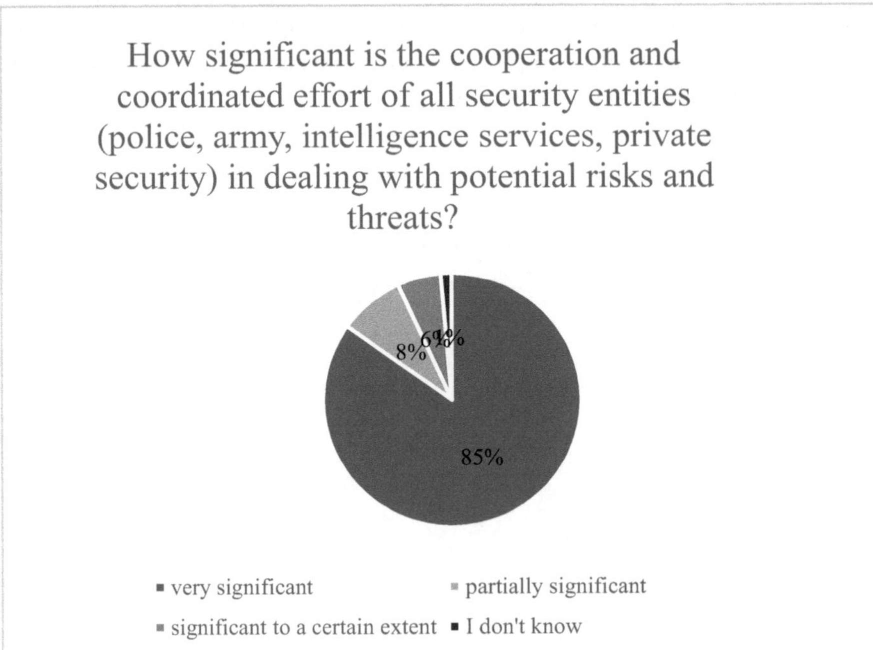

Consistently, in relation to the previously mentioned findings from the qualitative analysis of the normative-legal setting and the strategic provisions for defining the protection of critical infrastructure, the respondents who contributed with their opinions and views were asked the question that was created for the needs of the preparation of this paper, which reads: "Do you believe that it is necessary to adopt a legal framework for the protection of critical infrastructure?", which builds on the many years of efforts and initiatives by the stakeholders involved, to which the persons who gave their answer could choose: (1) yes to a large extent, (2) partially, (3) no or—(4) don't know.

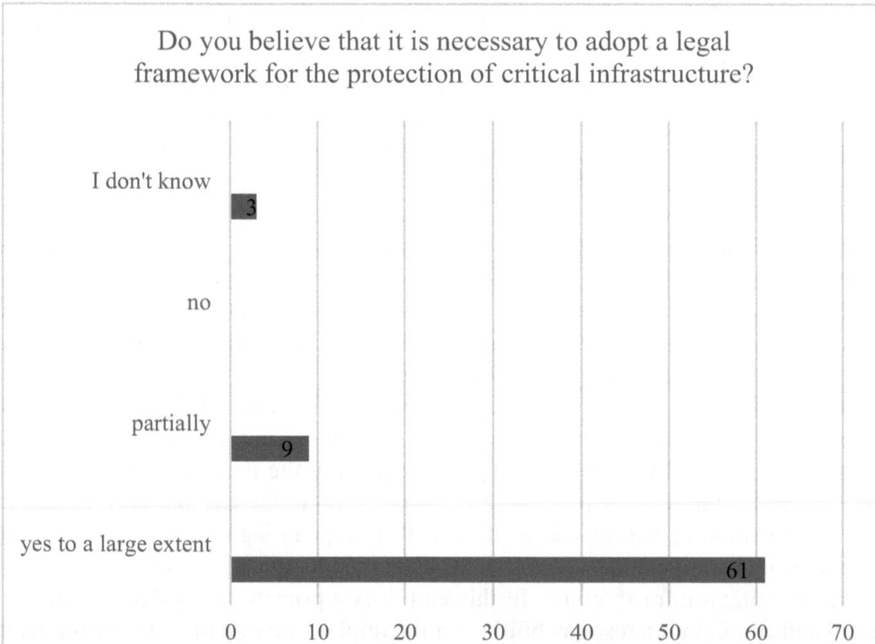

In terms of viewpoints, almost 84%, or 61 of the respondents, have an affirmative attitude towards the adoption of a law that defines the national critical infrastructure and regulates the instruments with which it is protected, i.e. it is largely necessary to define it within the framework of positive legal regulations, which is closely related to a number of other laws, in compliance with EU directives, a subject in a number of national and regional strategies, action plans, etc.

10.4 Discussion Instead of Conclusion

In the altered security environment, critical infrastructure is a legitimate object of protection, which must be recognized by the system as such through the imperative of the normative disposition. The trend of significant increase in attacks on critical infrastructure, as well as the domino effect of multiplying the magnitude of the consequences of potential attacks on elements of critical infrastructure systems, requires the unification and integration of efforts for their protection, both at national and supranational levels. In addition, attacks on critical infrastructure are becoming increasingly complex and sophisticated. For example, the combination of cyber and physical attacks poses a risk not only to the owners and operators of those assets, but also to their suppliers, customers, businesses, and individuals in the vicinity of the attacked asset, including adjacent and interconnected critical infrastructure elements that may be adversely affected. The damage caused by an attack on critical infrastructure systems has the potential for widespread, massive, and highly correlated consequences, impacting multiple sectors. Since the state occupies a central role in the national critical infrastructure protection system, it is in its greatest interest for it to operate smoothly, regardless of the ownership structure of a particular facility or network of critical infrastructure, thereby ensuring the smooth functioning of society.

From this perspective, it is necessary to raise awareness and a proper understanding of the significance of critical infrastructure within the strategic management of the state and its institutions, in order to optimize public and private cooperation and coordination through the competent ministries, regulatory bodies and entities that are direct owners or managers of critical infrastructure facilities to strengthen and improve critical infrstructure protection plans.

In order to improve national defense capacities and capabilities in accordance with strategic documents for fulfilling national defense commitments and obligations arising from NATO membership, strengthening the position of the Republic of North Macedonia in the process of approximation to the EU and harmonization and transposition of national legislation with European legislation, especially the EU Council Directive 2008/114/EC of 2008 on the identification and designation of European critical infrastructure. To this end, it is a priority to regulate the institutional and coordination responsibilities for the implementation of tasks arising from the area of critical infrastructure protection.

Taking into account the findings stated in this paper, it is unequivocally concluded that the protection of critical infrastructure, both from the aspect of normative-legal designation and from the aspect of operational action, extends through multiple legal provisions, by-laws, strategies and action plans whose bearers and proposers are different entities, from the state and private sectors.

From the perspective of the operational domain in the engagement of critical infrastructure protection, there is a general consensus that it is necessary to intensify the cohesiveness of the engagement in protection by all elements of the security and defense system.

Hence, the identification of critical infrastructure in North Macedonia counterpoises an imperative and is in line with the guidelines for normative regulation of issues in the field of identification, designation, analysis and protection of national and European critical infrastructure.

Finally, these are some of the conclusions that can be drawn from the analysis made according to the model of multi-criteria decisions in the legal, strategic and operational framework (MCDA) as a branch of operational research in the sciences of governance and management that involves the evaluation of a set of possible courses of action or options based on various quantitative and qualitative criteria.

References

1. Legal framework for securing critical infrastructure: with a focus on securing critical infrastructure in the Republic of North Macedonia.—Skopje: Chamber of the Republic of Macedonia for Private Security, pp 44 (2016)
2. Government of the Republic of Macedonia. Decision on amendments and supplementation of the decision on determination of legal entities that are obliged to have private security. Official Gazette of the Republic of Macedonia, No. 126 of 12.9.2017 (2017)
3. Pious RM (2001) The cuban missile crisis and the limits of crisis management. Polit Sci Q 116(1):81–105. https://doi.org/10.2307/2657821
4. The protection of critical infrastructures against terrorist attacks: compendium of good practices. United Nations office of counter-terrorism and interpol (2018)
5. Keković Z, Stanković N, Galić D (2023) The concept of resilience as a broader methodological approach for assessment of human security in local communities. Security Dialogues / Безбедносни дијалози 14:25–44
6. Critical infrastructure security: concepts and practices: a comprehensive guide to learn critical infrastructure security. Cybellium Ltd (2024)
7. Communication from the commission to the european parliament and the council on the seventh progress report on the implementation of the EU security union strategy. Brussels, 15.5.2024 https://eur-lex.europa.eu/resource.html?uri=cellar:c7ea112e-2248-11ef-a251-01aa75ed71a1.0024.02/DOC_1&format=PDF. Accessed 2025/01/06
8. NATO 2030 and the new strategic concept. NATO library https://natolibguides.info/nato2030/2022strategicconcept
9. CISA critical infrastructure sectors. https://www.cisa.gov/topics/critical-infrastructure-security-and-resilience/critical-infrastructure-sectors. Accessed 2025/06/03
10. Warsaw summit key decisions. NATO fact sheet (2017)
11. North Macedonia regulator adopts cybersecurity rules, takes on leading role in securing energy sector protection and resilience. https://pubs.naruc.org/pub/CAEE2894-9F59-5694-D501-A37BF40A6E54. Accessed 2025/06/01
12. Bakreski O, Milosevska T, Alceski G (2017) Critical infrastructure protection. Chamber of Private Security of the Republic of North Macedonia, Skopje
13. Roepke WD, Thankey H (2019) Resilience: the first line of defence. NATO Review (2019). https://www.nato.int/docu/review/articles/2019/02/27/resilience-the-first-line-of-defence/index.html. Accessed 2025/06/06, Pious RM (2001) The cuban missile crisis and the limits of crisis management. Polit Sci Q 116(1):81–105. https://doi.org/10.2307/2657821
14. Council directive 2008/114/EC of 8 december 2008 on the designation of European critical infrastructure and the assessment of the need to improve their protection. EU-Lex. European Union. https://eur-lex.europa.eu/eli/dir/2008/114/oj/eng. Accessed 2025/05/06
15. Cybersecurity strategy 2025–2028. Ministry of digital transformation. https://mdt.gov.mk/files/sodrzina_na_CCB_2025_2028_usoglasen_tekst.pdf. Accessed 2025/06/04
16. Defense strategy of the republic of North Macedonia. Ministry of defense of the Republic of North Macedonia (2020)
17. National convention on the European Union in the Republic of North Macedonia. Council Directive 2008/114/EC of 8 December 2008. https://nkeu.mk/2022/01/20/council-directive-2008-114-ec/. Accessed 2025/06/01
18. EU-NATO task force: final assessment report on strengthening our resilience and protecion of critical infrastructure https://ec.europa.eu/commission/presscorner/detail/it/ip_23_3564. Accessed 2025/06/04
19. Draft law on critical infrastructure. Ministry of defense of the Republic of North Macedonia. https://mod.gov.mk/storage/2022/12/predlog-Zakon-za-krivicna-infrastruktura.pdf. Accessed 19 May 2025
20. National development strategy 2024–2044. Government of the Republic of North Macedonia. https://nrs.mk/content/downloads/documents/thematic/Predlog-NRS%202024-2044%2026.09.2024.pdf. Accessed 2025/06/04

21. Resilience, civil preparedness and Article 3 NATO. https://www.nato.int/cps/en/natohq/topics_132722.htm. Accessed 2025/06/06
22. North macedonia accepted into NATO cyber defense center and in the process of full membership. Ministry of foreign affairs. https://mfa.gov.mk/en/page/13/post/3168/severna-makedonija-prifatena-vo-nato-sajber-centarot-za-odbrana-i-vo-procesot-na-polnopravno-chlenstvo. Accessed 2025/06/06
23. Kerigan-Kyro D (2014) NATO and critical infrastructure resilience—planning for the unknown. NATO science for peace and security series–E: human and societal dynamics. vol 116: Critical infrastructure protection, pp 1–12
24. Acquis communautaire (European Union Law). IPARD Dictionary (2025)
25. The national strategy for physical protection of critical infrastructures and key assets. The White House, Washington (2003)
26. European commission: EU-NATO task force: final assessment report on strengthening our resilience and protection of critical infrastructure. 29 June 2023 (2023)
27. Long-term plan for the development of defense capabilities 2023–2032. Ministry of defense of the Republic of North Macedonia. https://mod.gov.mk/storage/2023/10/DPROS-2023-2032-publikacija-za-web.pdf. Accessed 19 May 2025
28. Law on police (consolidated text) 2022. Ministry of interior of the Republic of North Macedonia. Skopje, April 2022 (2022)
29. Bakreski O, Milosevska T (2021) Information security and critical infrastructure. Directorate for security of classified information, Skopje (2021)
30. Documents in the MK/EU accession process. Parliament of the Republic of North Macedonia (2025)
31. Directive (EU) 2022/2557 of the European parliament and of the council of 14 december 2022 on the resilience of critical entities and repealing council directive 2008/114/EC (Text with EEA relevance)
32. Information on the process of drafting a draft Law on Critical Infrastructure. Ministry of Defense. (2022). https://mod.gov.mk/javna-rasprava-zakon-za-kriticna-infrastruktura/. Accessed 2025/08/06
33. Critical infrastructure security and protection. The public-private opportunity. White paper and guidelines by CoESS and its working committee critical infrastructure (2016)
34. Regulatory impact assessment report. Critical infrastructure act. Ministry of defense (2022)
35. Implementation of the law on private security and its impact in the Republic of Macedonia. Chamber of the Republic of Macedonia for Private Security, 01.06.2017
36. Law on private security (unofficial consolidated text). Ministry of internal affairs of the Republic of Macedonia. ("Official Gazette of the Republic of Macedonia" no. 166/12), https://mvr.gov.mk/zakon/3. Accessed 2025/05/12
37. Bellini E, D'Aniello G, Flammini F, Gaeta R (2025) Situation awareness for cyber resilience: a review. Int J CritAl Infrastruct Protc 49. https://doi.org/10.1016/j.ijcip.2025.100755
38. International security ligue. Critical infrastructure. Library. https://www.international-security-ligue.org/. Accessed 2025/06/08
39. Council directive 2008/114/EC of 8 december 2008 on the identification and designation of European critical infrastructures and the assessment of the need to improve their protection. EUR-Lex. European Union https://eur-lex.europa.eu/legal-content/EN/NIM/?uri=oj:JOL_2008_345_R_0075_01. Accessed 2025/05/06
40. Bakreski O, Gurovski M, Bardjieva ML (2021) Public and private security—coordination and cooperation. Chamber of the Republic of North Macedonia for Private Security, pp 160–162. Skopje (2021)

Chapter 11
CBRN Incidents and Critical Infrastructure: The Role of First Responders in an Integrated Emergency Management and Resilience Building in Serbia

Marina Dabetić, Zoran Keković, Milica Ćurčić, and Boris Obrovski

Abstract This paper investigates the perceptions of institutional capacities, preparedness levels, and interagency coordination among several organizations in the Republic of Serbia (RS) responsible for responding to chemical, biological, radiological, and nuclear (CBRN) incidents. The research, based on a quantitative survey conducted with 139 professionals from four major institutions, indicates a generally developed awareness of CBRN risks and the importance of effective early response. However, the findings also reveal differences in the depth of knowledge, availability of technical resources, and implementation of standardized procedures, which appear to stem from the specific mandates and operational experiences of individual institutions. While specialized units, such as the CBRN Center and the Sector for Emergency Management, demonstrate relatively advanced capacities and readiness, other agencies—such as the Customs Administration and Border Police—express the need for additional support in terms of targeted training and access to technical tools. Educational background and gender were also found to influence perceptions, as respondents with higher education levels and female respondents

M. Dabetić (✉) · M. Ćurčić
Department of Physical Chemistry, Vinča Institute of Nuclear Sciences–National Institute of the Republic of Serbia, University of Belgrade, Belgrade, Serbia
e-mail: fmarina@vin.bg.ac.rs

M. Ćurčić
e-mail: milica.curcic@vinca.rs

Z. Keković
Faculty of Security Studies, University of Belgrade, Belgrade, Serbia
e-mail: zkekovic@fb.bg.ac.rs

B. Obrovski
Department of Environmental Engineering and Occupational Safety and Health, Faculty of Technical Sciences, University of Novi Sad, Novi Sad, Serbia
e-mail: borisobrovski@uns.ac.rs

S. Kentera et al. (eds.), *Building Resilience in a Dynamic Critical Infrastructure Landscape*, NATO Science for Peace and Security Series C: Environmental Security,
https://doi.org/10.1007/978-3-032-04383-2_11

showed greater confidence in the existence of national-level CBRN response structures. Moreover, the results suggest that improving communication infrastructure and establishing mobile detection capabilities could significantly enhance the national response framework. Overall, the research emphasizes the importance of continued institutional investment in capacity-building efforts, interagency coordination mechanisms, joint training initiatives, and the promotion of gender equality in order to further harmonize and strengthen the national system for managing CBRN threats.

Keywords CBRN incidents · First response · Interagency coordination · Preparedness · Serbia

11.1 Introduction

The release of chemical, biological, radiological, or nuclear (CBRN) substances into the environment, whether accidental or caused by human intent [1–5], is of great concern to society. In addition to the ongoing threat posed by terrorists who are willing to use unconventional weapons to achieve their goals [1, 2], the risk of a CBRN incident is considered to be increased due to accelerated technological progress, the broader availability of certain types of CBRN substances, the prevalence of nuclear facilities and the chemical industry, medical and biological laboratories, as well as the widespread use of radioactive sources [6–8]. The main goal of first responders is to prevent a complete loss of control over the system by keeping its behavior and parameters within acceptable limits [9, 10]. First responders are the first to mitigate the threats and relieve the consequences of attacks [11]. Given the increasing complexity of modern threats, understanding how first responders perceive their responsibilities and capacities is essential for improving national response strategies and enhancing interagency coordination. A study was conducted in the Republic of Serbia (RS) with the aim of examining the perception and preparedness of first responder services in relation to CBRN risks, as well as their role within integrated emergency management and the strengthening of critical infrastructure resilience.

11.2 Methodology

A quantitative research design was applied in the study, with the aim of examining the perceptions of relevant institutions in the RS regarding their capacity to respond to chemical, biological, radiological, and nuclear (CBRN) incidents. The data collection method used was a survey conducted through a structured questionnaire, which was designed to cover several thematic units: organizational affiliation, first response, national-level infrastructure and resources, standardized procedures, and interagency coordination during initial response.

Table 11.1 Sociodemographic characteristics of the respondents

		Frequency (F)	Percent (P) [%]
Gender	Male	117	84.2
	Female	22	15.8
Level of education	High school	57	41
	Higher school	10	7.2
	Faculty	58	41.7
	Master studies	12	8.6
	Doctorate	2	1.4
Organizational unit	Customs administration	32	23
	CBRN center	22	15.8
	Sector of emergency situation	41	29.5
	Border police administration	44	31.7

11.2.1 Sample

The research included a total of 139 respondents, of which 117 were male and 22 female (Table 11.1). The youngest respondent was 22 years old, while the oldest was 64 ($M = 43.00$, $SD = 8.40$). Most respondents had completed either higher education (41.7%) or secondary education (41%), while the smallest number held a doctoral degree (1.4%) (Table 11.1). All respondents were employed, with years of work experience ranging from 1 to 44 years ($M = 18.72$, $SD = 8.86$). The participants were divided among four organizations: the Ministry of Finance – Customs Administration (23%), the CBRN Center (15.8%), the Sector for Emergency Management – Risk Management and Civil Protection Administration (29.5%), and the Border Police Directorate (31.7%) (Table 11.1).

Statistical Data Processing

Statistical data processing and analysis were conducted using IBM SPSS (Statistical Package of Social Science) software, version 25. Descriptive statistics and the chi-square test were applied to examine differences between groups. A significance level of 0.05 was used as the threshold.

11.3 Results

Based on the results presented in Table 11.2, it can be concluded that most respondents stated that their units are not the first to respond on the scene (73.7%). More than half of the respondents believe that at the national level of the Republic of Serbia, there exists a separate specialized service for receiving and forwarding information related to emergency response in the event of a CBRN agent release (54.5%). Many respondents believe that a national CBRN response team does not exist (50.8%). A large majority stated that their organizational unit does not have a mobile detection service for CBRN agents (73.8%). Furthermore, 59.7% believe that a national-level system for monitoring/early warning of potential CBRN incidents exists. Finally, a large majority of respondents indicated that they are not familiar with standardized procedures for inter-agency communication (67.2%) and procedures for operating in contaminated environments (63.6%).

The study investigated whether significant differences exist in respondents' answers based on their organizational unit. The Chi-square test was employed to assess these differences. The results indicate statistically significant differences across all questions in relation to the organizational unit (Table 11.3).

From the numerical and percentage distribution, it can be observed that primarily the Emergency Situations Sector responds first on the ground, whereas the Customs Administration and the CBRN Center never respond first.

Based on the numerical and percentage distribution, it can be noted that respondents from the Emergency Situations Sector and the Border Police Directorate predominantly believe that a specialized service/center exists at the national level for receiving and forwarding information related to emergency response needs, while respondents from the Customs Administration predominantly believe that such a service does not exist.

Similarly, perceptions about the presence of a CBRN response team differ significantly: respondents from the Emergency Situations Sector and Border Police

Table 11.2 Perception of first response and standardized procedures

		F	P [%]
Do members of your organizational unit respond first on the ground in case of escalation of CBRN risks? (P1)	Yes	36	26.3
	No	101	73.7
At the national level of the RS, is there a separate, specialized service/center for receiving and forwarding information regarding emergency response needs in case of CBRN agent release? (P2)	Yes	72	54.5
	No	60	45.5
At the national level of the Republic of Serbia, is there a CBRN response team for responding to escalation of CBRN risks? (P3)	Yes	62	49.2
	No	64	50.8
Does your organizational unit have mobile services for detection of CBRN agents? (P4)	Yes	33	26.2
	No	93	73.8
At the national level, is there a system for monitoring/early warning of potential CBRN incidents? (P5)	Yes	77	59.7
	No	52	40.3
Are you familiar with standardized communication procedures between services from different sectors? (P6)	Yes	43	32.8
	No	88	67.2
Are you familiar with standardized procedures for operating in contaminated environments with additional risks? (P7)	Yes	47	36.4
	No	82	63.6

Table 11.3 Differences in responses across organizational units

		Customs	Center CBRN	Emergency situations sector	Border police directorate	χ2	p
P1	Yes	0 (0%)	0 (0%)	25 (64.1%)	11 (25%)	48.088	< 0.0005*
	No	32 (100%)	22 (100%)	14 (35.9%)	33 (75%)		
P2	Yes	1 (3.1%)	10 (45.5%)	25 (73.5%)	36 (81.8%)	53.002	< 0.0005*
	No	31 (96.9%)	12 (54.5%)	9 (26.5%)	8 (18.2%)		
P3	Yes	1 (3.1%)	7 (35%)	21 (70%)	33 (75%)	45.705	< 0.0005*
	No	31 (96.9%)	13 (65%)	9 (30%)	11 (25%)		
P4	Yes	0 (0%)	15 (68.2%)	12 (42.9%)	6 (13.6%)	39.033	< 0.0005*
	No	32 (100%)	7 (31.8%)	16 (57.1%)	38 (86.4%)		
P5	Yes	1 (3.1%)	19 (90.5%)	24 (75%)	33 (75%)	58.229	< 0.0005*
	No	31 (96.9%)	2 (9.5%)	8 (25%)	11 (25%)		
P6	Yes	0 (0%)	11 (50%)	18 (52.9%)	14 (32.6%)	24.821	< 0.0005*
	No	32 (100%)	11 (50%)	16 (47.1%)	29 (67.4%)		
P7	Yes	1 (3.1%)	21 (95.5%)	17 (54.8%)	8 (18.2%)	59.283	< 0.0005*
	No	31 (96.9%)	1 (4.5%)	14 (45.2%)	36 (81.8%)		

* The Chi-square test was conducted at a significance level of 0.05

Directorate mostly confirm their existence, whereas those from the Customs Administration predominantly believe otherwise.

Regarding mobile detection units for CBRN agents, respondents from the CBRN Center generally affirm their presence, while those from the Customs Administration mostly deny it. Furthermore, most respondents from the Customs Administration consider that there is no system for monitoring or early warning of potential CBRN incidents, a view not shared by respondents from other units, who mostly confirm such a system's existence.

Familiarity with standardized inter-agency communication procedures also varies. Respondents from the Customs Administration generally report low awareness, whereas those from other units demonstrate partial familiarity. Lastly, knowledge of standardized operational procedures in contaminated environments is limited among respondents from the Customs Administration and Border Police Directorate, while respondents from the CBRN Center predominantly indicate good familiarity.

The study examined whether there are significant differences in respondents' answers based on their gender. To test for significant differences by gender, the chi-square test was applied. Based on the chi-square test results, a statistically significant difference was found for the question regarding the existence of a national-level CBRN response team for escalation of CBRN risks (Table 11.4). Numerical and percentage distributions indicate that female respondents predominantly believe that such a CBRN team exists, while male respondents predominantly believe that it does not.

Table 11.4 Differences based on respondents' gender

		Male	Female	χ2	p
P1	Yes	27 (23.3%)	9 (42.9%)	2.581	0.108
	No	89 (76.7%)	12 (57.1%)		
P2	Yes	59 (52.2%)	13 (68.4%)	1.132	0.287
	No	54 (47.8%)	6 (31.6%)		
P3	Yes	49 (44.5%)	13 (81.3%)	6.132	0.013*
	No	61 (55.5%)	3 (18.7%)		
P4	Yes	32 (28.6%)	1 (7.1%)	1.951	0.162
	No	80 (71.4%)	13 (92.9%)		
P5	Yes	66 (58.4%)	11 (68.8%)	0.267	0.605
	No	47 (41.6%)	5 (31.3%)		
P6	Yes	38 (33.3%)	5 (29.4%)	0.002	0.965
	No	76 (66.7%)	12 (70.6%)		
P7	Yes	44 (38.9%)	21 (18.8%)	1.672	0.196
	No	69 (61.1%)	1 (81.3%)		

The study examined whether respondents' answers significantly differ according to their level of education, using the chi-square test. The results indicate a statistically significant difference for all questions except question 6 (Table 11.5).

Analysis of the numerical and percentage distributions reveals that respondents with higher education are more likely to perceive their organizational unit as the first

Table 11.5 Differences based on level of education

		High school	College and university	χ2	p
P1	Yes	9 (15.8%)	27 (33.8%)	4.654	0.031*
	No	48 (84.2%)	53 (66.3%)		
P2	Yes	24 (42.1%)	48 (64%)	5.410	0.020*
	No	33 (57.9%)	27 (36%)		
P3	Yes	22 (38.6%)	40 (58%)	3.945	0.047*
	No	35 (61.4%)	29 (42%)		
P4	Yes	4 (7.1%)	29 (41.4%)	17.186	< 0.0005*
	No	52 (92.9%)	41 (58.6%)		
P5	Yes	23 (41.1%)	54 (74%)	12.922	< 0.0005*
	No	33 (58.9%)	19 (26%)		
P6	Yes	14 (24.6%)	29 (39.2%)	2.496	0.114
	No	43 (75.4%)	45 (60.8%)		
P7	Yes	6 (10.5%)	41 (56.9%)	27.628	< 0.0005*
	No	51 (89.5%)	31 (43.1%)		

to arrive at the site. Similarly, they tend to acknowledge the existence of a specialized service or center responsible for receiving and forwarding information.

Furthermore, respondents with higher education predominantly believe that a CBRN response team exists, whereas those with only a high school education generally consider that such a team is absent. A larger proportion of respondents with advanced education also recognize the presence of mobile detection services for HBRN agents.

Additionally, higher-educated respondents are more aware of the monitoring and early warning systems for potential CBRN incidents compared to their counterparts with secondary education. They also demonstrate greater familiarity with standardized procedures for operating in contaminated environments.

11.4 Discussion

The obtained result, which shows a low percentage of respondents stating that their service is the first to respond on the scene (73.7%), can be interpreted as an indicator that the concept of a first responder team in the CBRN context is not universally applied across all members of organizations. Instead, the initial response appears to rely on specialized and highly qualified teams composed of a small number of individuals from different sectors, often only one or two representatives per sector. This organizational structure implies that an adequate initial response depends on the level of training, readiness, and equipment of these teams instead of on the number. In this sense, first responders are organized in such a way that each team member contributes their specific expertise (e.g., medical response, detection, decontamination, security, etc.), which enables rapid and efficient intervention.

Research findings related to a service for receiving and forwarding information (54.5%) show that more than half of the respondents believe such a service exists. Respondents from the Emergency Situations Sector and the Border Police Directorate generally believe that such a service exists, while respondents from the Customs Administration mostly believe that it does not. This discrepancy may result from varying levels of awareness or involvement of these units in national CBRN risk management mechanisms. Organizations more directly engaged in emergency response are likely to have better insight into existing communication channels and processes, whereas the Customs Administration, whose activities are more focused on oversight and control at border crossings, may not have the same visibility within emergency response systems. However, a significant percentage still disagrees with this, which may indicate a lack of standardized intelligence and communication infrastructure among key actors. Therefore, there is a need for a centralized information-sharing system, as well as regular training sessions and simulation exercises involving all relevant organizations. The absence of mobile detection units within organizations (73.8%) may suggest that such equipment is held exclusively by specialized units or smaller groups that are part of designated first response teams. This implies that access to detection capabilities is not widely distributed across organizations but is limited to selecting well-trained professionals. Such differences in attitudes and perceptions between organizational units highlight the need for improved integration and communication within the CBRN risk management system.

The research results indicate significant differences in the level of awareness and perception of capacities and procedures among various organizational units, which may be a consequence of their differing roles, areas of focus, and degrees of involvement in CBRN incidents. These differences are particularly evident in responses to the question regarding deployment to the field in the event of a CBRN risk escalation and in the assessments of the existence of a specialized service for receiving and forwarding information. The results show that the Emergency Situations Sector most frequently responds first in the event of a CBRN risk escalation, while the Customs Administration and the CBRN Center never deploy first to the scene. These findings align with the expectation that the Emergency Situations Sector, as the main institution responsible for rapid response to various risks, serves as the frontline in such situations. On the other hand, the role of the Customs Administration and the CBRN Center is more focused on specific technical tasks and supports the initial response.

The CBRN Center, as a specialized unit, is more familiar with technical capacities and procedures, but may be less involved in broader coordination with other services. According to existing research, an effective, timely, and dependable response to CBRN incidents necessitates both well-equipped resources and highly trained personnel [11]. To improve the situation, it is necessary to enhance the training and education system, particularly within Customs and the Border Police, ensuring that members of all units are familiar with clear and standardized procedures. Furthermore, it is essential to strengthen coordination and communication among all relevant organizational units through joint exercises and experience sharing. This would help establish the necessary conditions for a more effective response to potential CBRN incidents, which is particularly important during the initial phase of response.

Furthermore, a statistically significant gender-based difference was observed in the perception of the existence of a national-level CBRN response team, with female respondents more likely to believe that such a team exists. This suggests possible variations in awareness or confidence in institutional capacities between male and female personnel. This finding may be partially explained by prior research indicating that women tend to exhibit specific characteristics in decision-making processes, such as analytical thinking, a consultative approach, honesty, and intuition [12], which may influence how they assess institutional structures and their effectiveness. Such traits could contribute to a greater inclination among women to recognize or place trust in the existence of formalized mechanisms for CBRN response.

These findings highlight the impact of educational background on the level of awareness and understanding of CBRN response mechanisms, indicating that higher education is associated with greater familiarity with key operational components and protocols. This is consistent with previous research, which has shown that higher risk perception is associated with higher levels of education, practical experience, and acute stress reactions [13].

Understanding these differences is essential for designing targeted training programs and improving preparedness strategies across diverse professional organizations responsible for first response.

The literature indicates a limited number of studies focusing on first responders in CBRN incidents, suggesting that future research should place greater emphasis on this group [9, 13].

11.5 Conclusion

The findings reveal disparities in the level of awareness of institutional capabilities, standardized procedures, and resource availability across different organizational units in RS, genders, and educational levels. Educational background proved to be a strong influencing factor, with more highly educated respondents showing greater awareness of national-level systems and procedures. Female respondents more frequently believed in the existence of formalized national CBRN structures, highlighting perceptual differences that may be relevant for future training approaches.

The results underline the importance of targeted capacity-building measures, including regular interagency training, improved access to detection and protection resources, and better dissemination of standardized operational procedures. Enhancing institutional coordination and investing in human and technical capacities are essential steps toward a more resilient and responsive CBRN emergency management system.

Future research should continue to explore the role of first responders in various national contexts, with a particular focus on organizational culture, perceived preparedness, and inter-institutional collaboration.

Acknowledgements This work was carried out within the scientific research activities of the 'Vinča' Institute of Nuclear Sciences - National Institute of the Republic of Serbia, funded by the Ministry of Science, Technology, and Innovation, grant number 451-03-136/2025-03/ 200017.

References

1. Sandström BE, Eriksson H, Norlander L, Thorstensson M, Cassel G (2014) Training of public health personnel in handling CBRN emergencies: a table-top exercise card concept. Environ Int 72:164–169
2. Koblentz GD (2020) Emerging technologies and the future of CBRN terrorism. Wash Q 43(2):177–196
3. Benolli F, Guidotti M, Bisogni F (2021) The CBRN threat. Perspective of an interagency response. In: International security management: new solutions to complexity, pp 429–448
4. Farhat H, Alinier G, Helou M, Galatis I, Bajow N, Jose D, Jouini S, Sezigen S, Hafi S, Mccabe S (2024) Perspectives on preparedness for chemical, biological, radiological, and nuclear threats

in the middle east and north Africa region: application of artificial intelligence techniques. Health Security
5. Karmon E (2017) Are we nearer to a major CBRN terrorism threat? In: Non-traditional security challenges in Asia, pp 240–252. Routledge, London
6. Ferguson CD, Kazi T, Bīrīrā J (2003) Commercial radioactive sources: surveying the security risks. Monterey Institute of International Studies, Center for Nonproliferation Studies
7. Carter H, Drury J, Amlôt R (2020) Recommendations for improving public engagement with pre-incident information materials for initial response to a chemical, biological, radiological or nuclear (CBRN) incident: a systematic review. Int J Disaster Risk Reduct 101796
8. Noschese P, Kostadinov R, MD AC, MD PN (2012) CBRN medical teams protection in case of disasters. G Med Mil 162(1):113–116
9. Chroust G, Rainer K, Sturm N, Roth M, Ziehesberger P (2011) Improving resilience of critical human systems in CBRN emergencies: challenges for first responders. Syst Res Behav Sci 28(5):476–490
10. Dabetić M, Janjić V (2023) Nuclear safety and nuclear security: conceptual definitions, legal aspects, and a holistic approach to the concepts. Vojno delo 75(4):34–45
11. Maciejewski P, Gawlik-Kobylińska M, Lebiedź J, Ostant W, Aydın D (2020) To survive in a CBRN hostile environment: application of cave automatic virtual environments in first responder training. In: Proceedings of the 3rd international conference on applications of intelligent systems, pp 1–5
12. Minasyan D, Tovmasyan G (2020) Gender differences in decision-making and leadership: evidence from Armenia. Bus Ethics Leader 4:6–16
13. Yücel H, Cengiz S (2020) Assessment of the relationship between the risk perception and preparedness attitudes of search and rescue and firefighting teams against event site hazards in CBRN events: Adana province case. Gümüşhane Üniversitesi Sosyal Bilimler Dergisi 11(Ek):347–354

Chapter 12
Digital Transformation in Pharmaceutical and Medical Device Regulation: Building Resilience During the COVID-19 Crisis in Serbia

Tatjana Stojadinović, Ilija Antović, and Marija Branković

Abstract The COVID-19 pandemic has posed significant challenges to regulatory frameworks globally, with pharmaceutical and medical device regulation being at the forefront of the crisis. This paper explores the role of digital transformation in strengthening the resilience of the pharmaceutical regulatory system in Serbia during the pandemic. By leveraging digital tools, the Serbian Medicines and Medical Devices Agency (ALIMS) was able to streamline regulatory processes, ensuring faster approval and distribution of critical medicines and medical devices. The study examines the technological adaptations made by ALIMS, including the integration of eServices and automation in decision-making, and assesses their impact on operational efficiency and the timeliness of regulatory responses. The findings highlight the transformative potential of digital solutions in enhancing regulatory capacity during health crises and offer insights into future improvements for building resilience in critical infrastructure. This paper contributes to the understanding of digital transformation in the regulatory context and provides a case study of how Serbia responded to the challenges of the pandemic.

Keywords Digital transformation · Pharmaceutical regulation · Medical device regulation · COVID-19 · Resilience · Serbia · ALIMS

T. Stojadinović (✉) · M. Branković
Medicines and Medical Devices Agency of Serbia, Belgrade, Serbia
e-mail: tatjana.stojadinovic@alims.gov.rs

M. Branković
e-mail: marija.brankovic@alims.gov.rs

I. Antović
Faculty of Organizational Sciences, University of Belgrade, Belgrade, Serbia
e-mail: ilija.antovic@fon.bg.ac.rs

S. Kentera et al. (eds.), *Building Resilience in a Dynamic Critical Infrastructure Landscape*, NATO Science for Peace and Security Series C: Environmental Security,
https://doi.org/10.1007/978-3-032-04383-2_12

12.1 Introduction

The COVID-19 pandemic has raised numerous questions and challenges across all sectors of society. In this context, the healthcare system became a priority as the first line of defense against the virus and its far-reaching consequences on other areas of society [1]. For the healthcare system to function effectively during the pandemic, it was essential for all its components to operate not only smoothly but with greater efficiency than under normal circumstances. The disruptions caused by the pandemic were particularly evident in the drastic changes in demand for certain medicines, but most notably in the demand for specific medical devices [2]. To ensure the uninterrupted supply of essential medical products in deficit, it became necessary to facilitate the circulation of domestically produced medical devices or to enable the urgent importation of medicines and medical devices not yet registered in Serbia. Since a rapid response was crucial to combating the virus, as well as ensuring that medicines and medical devices put into circulation were effective, safe, and of acceptable quality, it was critical that the processes related to the regulation of medicines and medical devices function efficiently during the crisis. In this way, the national regulatory agency for medicines and medical devices played a crucial role in overcoming the crisis.

Different countries around the world reacted in various ways to the outbreak of the pandemic [3, 4] and to a greater or lesser extent used the advantages of digital technologies in different fields [5]. A positive example is Taiwan, where a system was developed within 72 h to combat the spread of the disease by utilizing data from existing healthcare databases and linking them with citizens' travel data and disease symptom information. By using Big Data analytics and mobile computing concepts, early identification of potential hotspots and timely response were enabled [6].

The introduction of emergency measures in the Republic of Serbia, including movement restrictions, necessitated changes in the usual business operations. Different industries adapted to the new circumstances in various ways, but what was common to all was the focus on building resilience by providing services electronically, using electronic business solutions.

12.1.1 Digital Transformation in Pharmaceutical Regulation

Digital transformation refers to the process of altering an organization's business processes to enable efficient operation through digital means [7].

It involves the integration of digital technologies across all business areas, leading to fundamental changes in operations and value delivery, radically transforming business models, generating new revenue streams, and completely altering organizational processes. This transformation brings challenges in leadership, encourages new ways of thinking, fosters innovation, and requires constant adaptation [8].

In pharmaceutical regulation, digital transformation is a key component of eGovernment in Serbia. The implementation of e-business solutions in pharmaceutical regulation, eHealth, and ePharmacy enhances communication, document management, and record-keeping in public administration, the pharmaceutical industry, and the healthcare system, while generating measurable savings in terms of time and cost [9].

To ensure faster and better access to medicines and medical devices for patients and healthcare professionals in Serbia, ALIMS has worked to improve transparency and efficiency in its operations with the pharmaceutical industry. Over recent years, ALIMS has undertaken extensive activities to enhance electronic business, digitalization, and data transparency. By adopting a digital business model, ALIMS has shifted its focus towards faster request processing, business analytics, and improved operational efficiency.

Through digitalization and the opening of data sets on the national open data portal (data.gov.rs), ALIMS has enabled healthcare organizations, government agencies, and the pharmaceutical industry to fully automate their processes. Most importantly, this initiative has contributed to improving the standard of living and quality of life for the citizens of the Republic of Serbia. ALIMS was one of the first agencies to adopt the concept of "open data" back in December 2015 [10].

In mid-2018, activities were initiated to implement a new modern information system for ALIMS, based on the digital transformation of all business processes, with the goal of increasing the application of electronic business and digitalization. This system developed new e-services that cover the full range of ALIMS operations. The new information system, based on eServices, represents the highest level of sophistication according to the Strategy for the Development of E-Government in the Republic of Serbia [11, 12]. The system is integrated with the eGovernment portal (euprava.gov.rs), enabling ALIMS clients, particularly the pharmaceutical industry in Serbia, to electronically submit all requests with the accompanying documentation in digital format, as well as facilitate two-way interaction and information exchange.

12.1.2 eServices in COVID-19 Pandemic

The provision of electronic services during the crisis has been carried out in various ways across different sectors [5]. For example, educational institutions have transitioned to distance learning, utilizing platforms like Moodle et al. [13]. In healthcare, the concept of telemedicine has been known for a long time but is still underused in Serbia [14]. This slower adoption of digital transformation in healthcare can be attributed not only to technical barriers but also to perceptions about the suitability of healthcare services for digitalization, including ethical concerns [15]. However, parts of the healthcare system, particularly in pharmaceutical regulation, have successfully transitioned to electronic services. Pharmaceutical regulations ensure the safety, efficacy, and quality of medicines through legal, administrative, and technical measures [16].

In a crisis, it is crucial to ensure the availability of effective, safe, and high-quality medicines and medical devices. To expedite the process of placing medicines and devices on the market, the digital transformation of regulatory processes in national agencies, such as ALIMS, is essential. Following the declaration of the state of emergency in Serbia and the recommendations of the government, many activities were accelerated, necessitating the submission of documentation electronically. This further facilitated the application of digital transformation in regulatory processes for medicines.

In Serbia, legal frameworks specifically address the regulation of medicines and medical devices in emergencies, allowing for the modification of existing procedures to prevent severe public health consequences. These provisions can be found in the Medicines and Medical Devices Law, Articles 50 and 206 [17], and the Medical Devices Law, Article 10 [18]. These regulations are designed to adapt pharmaceutical and medical device regulation to the unique needs of an emergency situation. For example, during the crisis, the issuance of expert opinions on the import of masks and gloves for internal use was facilitated electronically via the ALIMS portal.

The ultimate goal of this paper is to assess how digital transformation in pharmaceutical and medical device regulation has enhanced the resilience and efficiency of these processes during the COVID-19 pandemic and in normal circumstances. The results of this analysis can serve as a methodological framework for building resilience and ensuring effective responses in current and future crises, applicable not only in healthcare but also in other sectors.

12.2 Methods

This section will outline the stages through which ALIMS'operations have undergone digital transformation, followed by an explanation of how the impact of this transformation on ALIMS'operations will be measured.

12.2.1 Phases of the Digital Transformation of ALIMS' Business Processes

The digital transformation of ALIMS'business processes took place in four phases:

1. The first phase can be characterized as the phase of data openness, which began in 2015. Open data was made available on the National Open Data Portal of the Republic of Serbia.
2. In the second phase, starting in 2016, the submission of requests for services within ALIMS'operations was enabled electronically through the state eGovernment portal of the Republic of Serbia—eUprava (euprava.gov.rs).

3. On December 2, 2018, a new information system for medicines and medical devices was introduced, based on the digital transformation of business processes. This system allowed for the evaluation of items and the issuance of permits for medicines and medical devices by developing eServices of the highest level of sophistication, including two-way communication with clients. It was integrated with the eUprava portal. The system is based on web technology and ALIMS portals: lekovi.alims.gov.rs and ms.alims.gov.rs, which are used by both external users (pharmaceutical industry) and internal users (ALIMS evaluators).
4. The fourth phase was related to the adaptation of business processes, as well as the existing information system, to the crisis situation caused by the COVID-19 pandemic and the state of emergency declared in mid-March 2020. This resulted in the acceleration of the digital transformation process.

ALIMS began its digital transformation in 2015 with participation in the Serbian Government's Open Data project. By late 2018, ALIMS, in collaboration with the Institute of Public Health of Serbia, received the Open Data Champion award from the United Nations Development Programme (UNDP) and the Serbian e-Government Office.

The "eSubmission" project, launched in mid-2016, aimed to enable electronic submission of requests via the eGovernment portal. By 2017, 60 eServices for human and veterinary medicines were successfully implemented, continually updated to meet industry needs. This electronic system allows the pharmaceutical industry to operate without the need for physical visits to ALIMS, integrating the pharmaceutical regulatory subsystem into Serbia's broader e-government framework.

In June 2018, ALIMS launched a new information system for medical devices, named ADIS (Agency Digital Information System), which was fully operational by December 2018. This system enables permit holders to submit requests electronically via the ms.alims.gov.rs portal, while ALIMS evaluators process and monitor the status of submitted requests. Parallel development occurred for the digital transformation of pharmaceutical processes for human and veterinary medicines.

The implemented information system, based on advanced eServices, facilitates two-way interaction between external users (applicants) and ALIMS internal users (evaluators). It also includes online transactions, document verification, and proactive eServices. Following the introduction of emergency measures, the number of electronic services grew to 94, covering both medicines (70) and medical devices (24).

The emergency situation accelerated digital transformation beyond ALIMS, including processes with other government agencies. A key example is the streamlined process for importing medicines and medical devices, where ALIMS and the Customs Administration established an electronic system for collaboration, replacing paper documentation. This shift expedited customs procedures for essential medical supplies during the pandemic, ensuring timely delivery to the healthcare system and patients.

12.2.2 Measuring the Impact of Digital Transformation on Business Process Improvement

To measure the impact of digital transformation on ALIMS' business improvement, the analysis will first focus on the duration of processes related to the registration of medical devices (adding, modifying, and deleting records). Since managing and maintaining the medical device registry is a core function of ALIMS, and the highest volume of requests is for services related to these processes, it is essential to evaluate how digital transformation has affected this process. The impact will be assessed by examining the reduction in the number of days required to process requests after the introduction of the new information system, starting from December 2018.

The new Law on Medical Devices has significantly reduced processing times from 90 to 30 days, and this analysis will show how the digital transformation aligned with the new law, accelerating the process despite the number of staff remaining the same and the number of requests increasing. Additionally, changes in processing speed during the emergency situation beginning in March 2020 will be examined, as legal adjustments during this period led to further digital transformation of the process.

Apart from ALIMS'core processes, special attention will be given to services and business processes that became critical during the emergency. These processes, most often used during the crisis, can be identified by increased demand for specific services. This analysis will examine the change in demand for these services over the course of at least one year to highlight the impact of the ongoing crisis versus seasonal fluctuations. A detailed comparison of the request resolution times before, during, and after the digital transformation will be provided for each identified critical service.

The analysis will utilize open data published by ALIMS on the Republic of Serbia's Open Data Portal, cross-referenced with integrated data from ALIMS'old and new information systems. The analysis will address two key aspects: time taken to resolve requests and the volume of requests for specific ALIMS services.

The first aspect focuses on the change in the time required to process requests before and after the digital transformation, as well as during the crisis period. As previously mentioned, digital transformation has enabled the majority of data used in the request resolution process to be stored and processed electronically rather than on paper. Additionally, data in electronic form is exchanged between the different parties involved in the process, from the moment the request is submitted through the eGovernment portal or ALIMS portals, to the issuance of the decision.

To analyze the change in the speed of request resolution, the following methodology will be applied:

1. Data Selection: All requests for which a decision was issued by the end of April 2020 will be considered. The number of days taken from the moment the request was submitted to the issuance of the decision will be measured.
2. Reference Values: For each analyzed month within the period, the average number of days required to resolve requests will serve as a reference value. This

includes data from both before and after the implementation of digital transformation, and will also encompass requests processed during the crisis period (from mid-March to the end of April 2020, until the data required for analysis were available).
3. Graphical Representation: Data will be presented in graphs to visually highlight changes in the time taken to resolve requests. These graphs will be accompanied by the date when the digital transformation of each specific service was introduced, allowing for a direct comparison between the change in resolution time and the implementation of the digital transformation.
4. Focus on Registration Process: The analysis will specifically focus on the request data from the medical device registration process for human use (adding, modifying, and deleting records from the registry), as this is a representative and crucial process for ALIMS. The registration process is critical as it handles the majority of the agency's requests, particularly those related to regulatory changes.

Additionally, particular attention will be given to requests processed during the crisis situation, as the number of requests for critical services significantly increased from the beginning of the crisis. The increase in requests during the crisis is essential to understanding the impact of digital transformation during emergency situations. By analyzing data from this period, we can determine whether the digital transformation has contributed to faster processing times despite the increased number of requests.

The second aspect focuses on the change in the number of submitted requests for services before and after the digital transformation, as well as during the crisis period. The digital transformation of ALIMS' business processes, the changes in ALIMS' information system, and its integration with the eUprava portal, along with the specialized ALIMS portals, have made services more accessible to clients by allowing them to be submitted electronically. As a result of digital transformation, there has been an increase in the number of requests for services that are now electronically available. Therefore, the analysis of the change in the number of submitted requests will follow the methodology outlined below:

1. Data Selection: The analysis will focus on requests for which decisions were issued up to end of April 2020. The total number of submitted requests will be calculated for each month within the analyzed period.
2. Reference Values: For each Month, the Total Number of Requests Submitted Will Be Presented to Provide a Clear Understanding of how the Volume of Requests Has Changed over Time. Data Will Be Shown in Graphical Form to Make the Changes Easily Identifiable
3. Graphical Representation: The data will be visualized in graphs, which will include a comparison of the number of requests before and after digital transformation, and during the crisis period. For each service analyzed, the date when the digital transformation was implemented will be displayed, establishing a direct connection between the change in the number of requests and the digital transformation.
4. Impact of the Crisis Situation: Special attention will be given to the increase in the number of requests for critical services during the crisis. The data for March

and April 2020 will be compared to the same period in the previous two years to highlight any significant changes in demand that are a result of the crisis, rather than seasonal fluctuations.
5. Critical Services: For each identified critical service during the crisis, the change in the number of requests will be compared to the change in resolution times. This will be done under the assumption that the number of employees working on processing requests remained the same throughout the analysis period.
6. Exclusion of Non-Critical Processes: While the medical device registration process for human use is significant, the data on changes in the number of requests for this process will not be used as an objective indicator of the effects of digital transformation during the crisis, as the process was limited to urgent requests only due to the emergency measures.

This approach will allow for a comprehensive analysis of the increase in demand for critical services during the crisis, as well as the relationship between the digital transformation and the increase in request volume. The analysis will show how the increased demand during the crisis impacted the efficiency of the system and how digital transformation helped to accommodate this higher volume of requests.

12.3 Results

12.3.1 Analysis of the Impact of Digital Transformation on ALIMS' Core Process

At the beginning of this section, data related to the process of Registration of Medical Devices will be presented, which includes several different administrative procedures: registration of medical devices, renewal of registration, modification and supplementation of medical devices, and extension of registration. The first aspect of analyzing the effects of digital transformation—the change in the time required to process requests before and after the digital transformation, as well as during the crisis period—will be illustrated in the following diagram (see Fig. 12.1). The diagram covers resolved requests for all the services related to this process, with submission dates from January 1, 2018, to April 15, 2020. For each month, the average number of days taken to resolve requests submitted in that month will be calculated. It is important to note that the digital transformation of the services, represented in the diagram, was implemented on December 2, 2018, at which point requests have been fully processed electronically.

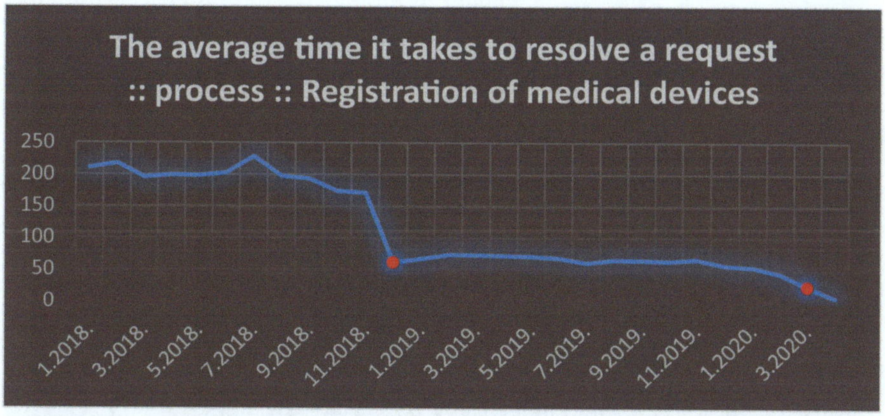

Fig. 12.1 Change in the time required to resolve the request—registration of medical devices

12.3.2 Analysis of the Impact of Digital Transformation on Critical Processes During the Crisis Situation

During the state of emergency, an analysis of the number of requests for all eServices related to medicines and medical devices revealed that there was particularly high interest in eServices related to expert opinions for the import of masks and gloves, which were most needed during the COVID-19 pandemic, as well as import permits for medical devices that were not registered. As a result, these processes will be treated as critical during this crisis, and the analysis of the effects of digital transformation on these processes will cover both aspects: the change in the time required to process requests and the change in the number of requests submitted before and after the digital transformation, as well as during the crisis period.

The first critical process to be analyzed is the process of Expert Opinions. This process can relate to various matters, but during the crisis, it became particularly significant in terms of determining whether a specific product is classified as a medical device, particularly for the import of masks and gloves. The increased demand for these medical supplies has been observed globally, in addition to Serbia [19].

Figure 12.2 shows the change in the time required to resolve requests for Expert Opinions before and after the digital transformation in December 2018, as well as after further adjustments to the process due to legal regulation changes at the onset of the COVID-19 crisis. The data includes requests resolved by the end of April 2020. Figure 12.2 provides a comparative view of the average resolution time for Expert Opinions requests submitted in the first four months of 2018—before digital transformation, then the same period in 2019—after digital transformation, and finally in the same period of 2020—where we can track the period before the crisis (January, February, and part of March) and after the crisis began (part of March and April).

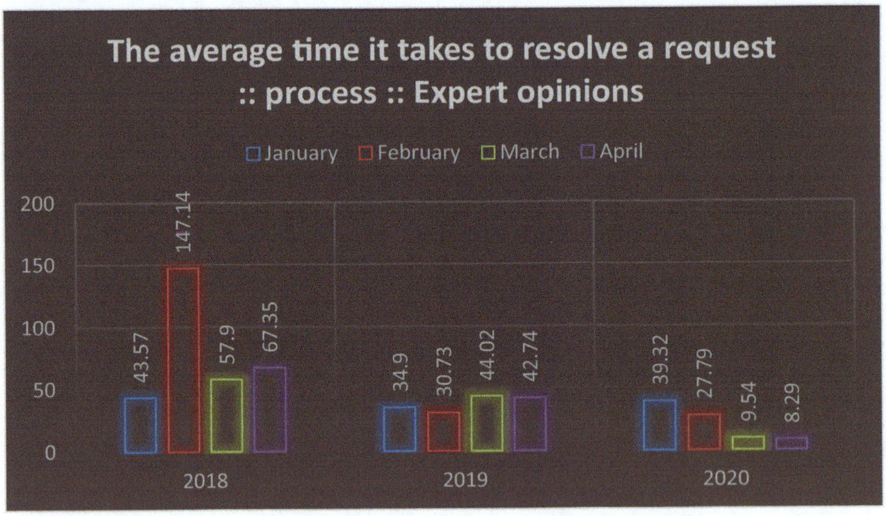

Fig. 12.2 Change in processing time for requests—expert opinions

In addition, Fig. 12.3 provides a comparative view of the number of requests submitted for Expert Opinions during the specified time periods.

Figure 12.4 illustrates the relationship between the change in the average number of days required to resolve requests for Expert Opinions and the change in the number of requests submitted for the same process during 2020, with the note that the crisis situation began in mid-March 2020.

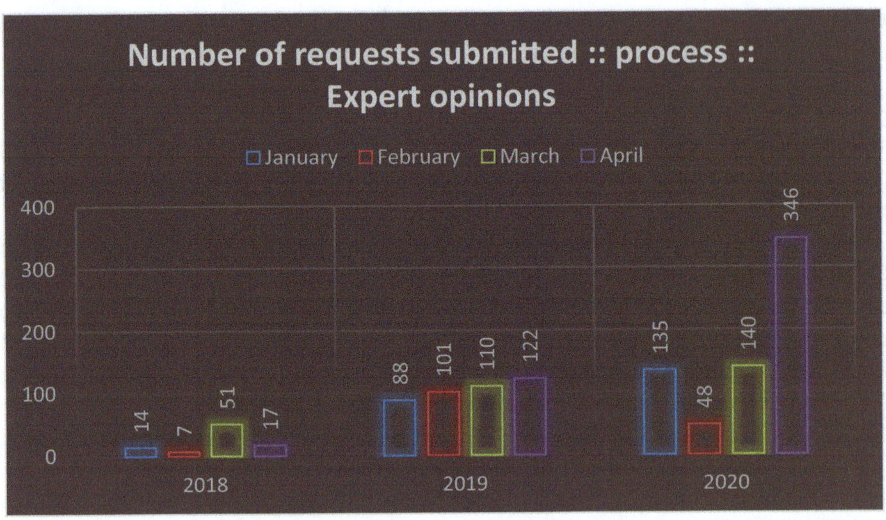

Fig. 12.3 Change in the number of requests—expert opinions

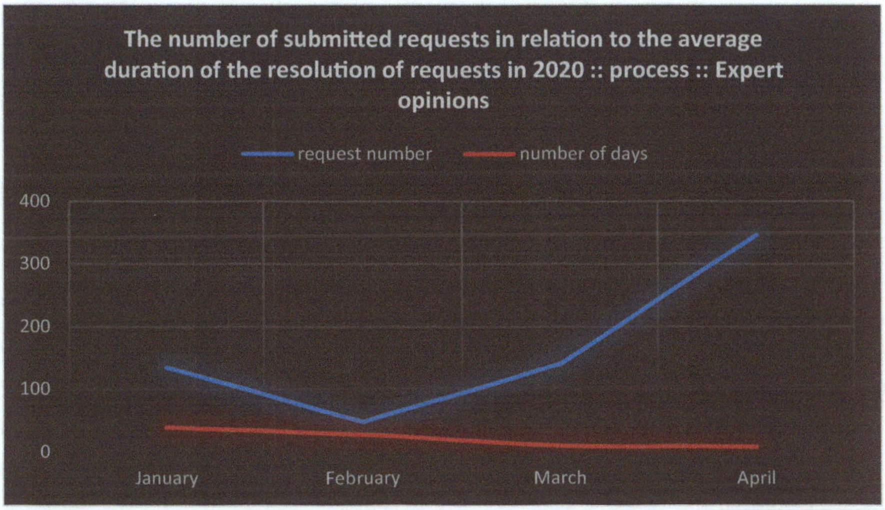

Fig. 12.4 Number of requests for expert opinions in relation to processing speed in 2020

The second critical process to be analyzed during the crisis situation is the process of Import Permits for Medical Devices Not Registered in Serbia. The agency can exceptionally approve the importation of a medical device that is not registered in the Republic of Serbia if it is intended for a specific patient or group of patients, for donation or humanitarian aid, as part of a donation program in the European Union, or for scientific research purposes, as well as in the case of an emergency situation in accordance with the law.

Figure 12.5 shows the change in the time required to process requests for Import Permits for Medical Devices Not Registered in Serbia before and after the digital transformation in December 2018, as well as after further adjustments to the process due to legal regulation changes at the onset of the COVID-19 crisis. The data includes requests resolved by the end of April 2020. Figure 12.5 presents a comparative view of the average resolution time for Import Permits for Medical Devices Not Registered in Serbia requests submitted in the first four months of 2018—before digital transformation, then in 2019—after digital transformation, and finally in the same period in 2020—where the period before the crisis (January, February, and part of March) and after the crisis began (part of March and April) can be compared.

In addition, Fig. 12.6 provides a comparative view of the number of requests submitted for Import Permits for Medical Devices Not Registered in Serbia during the specified time periods.

Figure 12.7 illustrates the relationship between the change in the average number of days required to resolve requests for Import Permits for Medical Devices Not Registered in Serbia and the change in the number of requests submitted for the same process during 2020, with the note that the crisis situation began in mid-March.

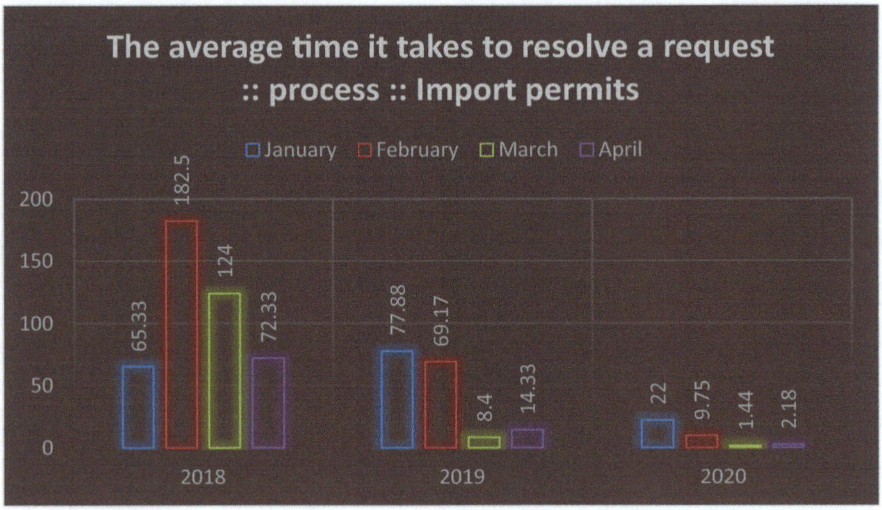

Fig. 12.5 Change in processing time for requests—import permits for medical devices not registered in Serbia

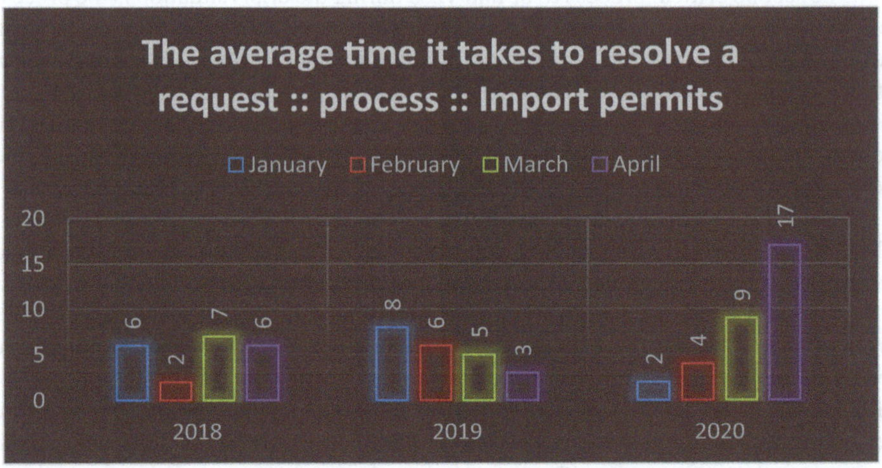

Fig. 12.6 Change in the number of requests—import permits for medical devices not registered in Serbia

12.4 Discussion

Since its establishment in 2004, the Medicines and Medical Devices Agency of Serbia (ALIMS) faced a recurring issue related to meeting regulatory deadlines for medical device registration, which was critical for maintaining public health safety. In 2018, the adoption of the new Law on Medical Devices reduced the processing

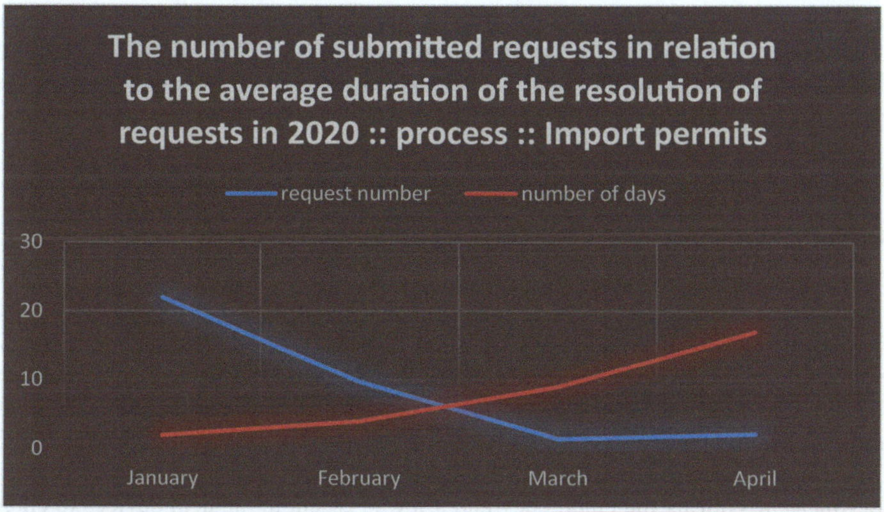

Fig. 12.7 Number of requests for import permits for medical devices not registered in Serbia in relation to processing speed in 2020

time for key business processes, such as medical device registration, from 90 days to 30 days. The only solution to comply with these deadlines was the digitalization, or digital transformation, of ALIMS' business processes and the development of a new information system based on electronic submission and processing of requests and accompanying documentation.

12.4.1 Impact on Processing Time

Analysis of the results indicates significant benefits achieved through the new information system, based on the digital transformation of business processes for medical devices. This transformation was evident in the substantial reduction in the number of days required to process, evaluate, and resolve cases related to medical device registration and issue permits (see Fig. 12.1). Prior to December 2018, the average processing time was about 199 days (with requests in July 2018 taking as long as 228 days). However, following the digital transformation in December 2018, a noticeable improvement was observed. In 2019, the average processing time dropped to 67 days, which was three times shorter. This trend continued, with the average number of days decreasing to just 32 days in 2020, more than six times shorter than the pre-transformation period. It is important to note that this average also includes the clock stop period, when the applicant needs to submit additional documentation for the request to be complete and ready for evaluation, which does not count towards the 30-day deadline.

With the new information system in place, the processing of requests related to medical device registration became significantly faster. This allowed ALIMS to meet legal deadlines, fulfill client needs, and, most importantly, timely ensure the protection of public health.

12.4.2 Impact of the COVID-19 Crisis on Medical Device Registration

During the COVID-19 emergency situation, the need for medical device registration decreased. This was because the registration process and the preparation of the required documentation became more time-consuming, and in the context of a public health emergency, such processes were difficult to maintain. The demand for medical devices surged. To address this, the government relaxed certain regulatory measures during the emergency and allowed essential medical devices to enter the market through expert opinions for the import of masks and gloves and import permits via donations. Consequently, the number of requests for Expert Opinions and Import Permits increased, and the time required to resolve these requests was reduced, thanks to the implementation of digital transformation.

In 2018, the processing time for Expert Opinions ranged from 43.57 days in January to 147.14 days in February. By 2020, the processing time for Expert Opinions dropped significantly, with an average of 9.54 days in March and 8.29 days in April during the COVID-19 pandemic. A significant decrease in processing time was also observed in 2019 (with an average of 38 days), compared to 2018 (with an average of 78 days) when the requests were processed in the old system (see Fig. 12.2).

12.4.3 Increasing Demand for Expert Opinions

The number of requests for Expert Opinions increased significantly from 2018 to 2020, reflecting the growing demand for medical supplies during the pandemic. In January 2019, there were 88 requests, rising to 122 in April 2019. In January 2020, there were 135 requests, with a notable surge to 140 requests in March 2020 and 346 in April 2020, demonstrating the increased demand during the pandemic (see Fig. 12.3).

12.4.4 Impact on Import Permits

Similarly, the Import Permits for Medical Devices process saw a significant rise in requests during the crisis. In April 2020, the number of requests for import permits increased to 17, compared to the pre-crisis period, when there were typically around 6–7 requests (see Fig. 12.6). The digital transformation of these business processes also resulted in a drastic reduction in processing time. In April 2020, the average processing time for import permits was reduced to about 2 days, compared to 182 days in February 2018 (see Fig. 12.5).

12.4.5 Resilience in Regulatory Processes

The impact of digital transformation in the Expert Opinions and Import Permits processes is particularly evident when considering the COVID-19 crisis. During the pandemic, the number of requests for both processes significantly increased, yet the time required to process these requests was reduced, despite the fact that the number of employees working on resolving requests remained unchanged (see Figs. 12.4 and 12.7). This demonstrates how digital transformation enhanced resilience in ALIMS' regulatory processes, enabling the agency to meet the surging demand while maintaining efficiency.

12.5 Conclusion

This paper demonstrates how societal changes can drive organizational transformation, including business process reconfiguration and the complete transformation of company operations. Special emphasis was placed on the changes related to the crisis situation, where established regulatory frameworks—specifically those governing pharmaceuticals and medical devices—had to be adjusted to enable the market to be quickly supplied with essential medicines and medical devices to ensure public health protection.

The aim of the digitalization and digital transformation of pharmaceutical and medical device regulatory processes is to increase efficiency and service quality through the integration of eGovernment, eHealth, and ePharmacy, resulting in overall benefits for the state, industry, and citizens. In this way, cooperation between ALIMS, the Ministry of Health, other government institutions, the pharmaceutical industry, healthcare institutions, and citizens has been enhanced through the use of open data on the national portal (data.gov.rs) and high-level eServices on the ALIMS portals (ms.alims.gov.rs, lekovi.alims.gov.rs) and eGovernment (euprava.gov.rs).

Digital transformation in pharmaceutical and medical device regulation has significantly accelerated the availability of high-quality, efficient, and safe medicines and medical devices to patients and healthcare professionals. It has also increased operational efficiency and effectiveness by reducing costs for market participants and

ALIMS itself. Processing times have been reduced, and the need for physical presence at ALIMS offices has been eliminated in favor of completely electronic procedures. This has proven to be of enormous importance for the efficient functioning of the healthcare system, especially during the pandemic.

By identifying the phases through which ALIMS' operations were digitally transformed, and analyzing the results that unequivocally show the positive impact of digital transformation on the efficiency of the observed processes, a methodological framework has been established for effectively responding in this and any future crisis. This framework is applicable not only in the healthcare system but also in other sectors.

Through this transformation, resilience has been built into regulatory processes, ensuring that ALIMS can quickly adapt to dynamic and unforeseen challenges, such as public health emergencies, and ensuring that the system remains operational and efficient, even in times of crisis. The continued development of eServices lays the groundwork for sustainable growth, better public service delivery, and enhanced collaboration across sectors, strengthening the resilience of Serbia's regulatory landscape well beyond the pandemic.

References

1. Commission on a global health risk framework for the future; national academy of medicine, Secretariat. The neglected dimension of global security: a framework to counter infectious disease crises. National Academies Press (US), Washington (DC) (2016)
2. Mircheva RD (2020) A crisis within the crisis: the impact of Covid-19 and Brexit on supply chains in the pharmaceutical industry. J Varna Univ Econ 3:352. Department of International Economic Relations, University of Economics, Varna, Bulgaria, Izvestiya
3. Bento F, Couto KC (2021) A behavioral perspective on community resilience during the COVID-19 pandemic: the case of Paraisopolis in Sao Paulo, Brazil. Sustainability 13:1447
4. Jon C, Kai K (2020) Countries test tactics in 'war' against COVID-19. Science 367:1287–1888
5. Andrea A, Nathalie C, Chiara F, Gloria F, Rocco F (2021) Distance learning as a resilience strategy during Covid-19: an analysis of the Italian context. Sustainability 13:1388
6. Wang CJ, Ng CY, Brook RH (2020) Response to COVID-19 in Taiwan: big data analytics, new technology, and proactive testing. JAMA 323(14):1341–1342
7. Leimstoll U, Dannecker A, Knechtli H, Quade M, Tanner C, Wölfle R (2018) (2018) E-business in the era of digital transformation. In: Dornberger R (ed) Business information systems and technology 4.0. studies in systems, decision and control, vol 141. Springer, Cham
8. Sofronijević A, Milićević V, Ilić B (2017) Strategijski pristup digitalnoj transformaciji poslovanja. Tehnika-Menadzment 67:273–279
9. Tatjana S, Vesela R, Božidar R (2010) E-business in the regulation of medicines in Serbia. Drug Information J 44(2):177–187. Drug Information Association, Inc. Printed in USA. ISBN 0092-8615
10. Tatjana S, Ljiljana R (2018) Banka otvorenih podataka u funkciji razvoja e-zdravstva. In: 17th international symposium INFOTEH-JAHORINA; 21–23 Mar 2018
11. European Commission (2010) Method paper 2010—preparing the 9th benchmark measurement, prepared by Capgemini, Rand Europe, IDC, Sogeti and DTi for the European Commission. Accessed 20.02.2021 from https://joinup.ec.europa.eu/collection/egovernment/document/eu-method-paper-2010-preparing-9th-benchmark-measurement

12. Strategija razvoja elektronske uprave u Republici Srbiji za period od 2015–2018. Godine i akcioni plan za sprovođenje strategije za period 2015–2016. In: godine, Sl. glasnik RS, br. 107/2015, Belgrade, Serbia (2015)
13. Vita H, Serhiy S, Yaroslav S (2020) How learning technology wins coronavirus. ICHTML 2020, SHS web of conferences 75. https://doi.org/10.1051/shsconf/20207500001
14. Volha S (2016) Sumary of the results of the study: potential benefits of the application of telemedicine in Serbia, based on master thesis. American Public University
15. Marialuisa S, Clara B, Paolo P, Primiano DN, Mattia L (2018) Cotton and surgical masks—what ecological factors are relevant for their sustainability? Sustainability 10:3548. 10.3390/su10103548, www.mdpi.com/journal/sustainability
16. Al-Worafi YM (2020) Drug safety in developing countries (Chapter 3—Medications registration and marketing: safety-related issues). Academic Press, pp 21–28. ISBN 9780128198377. https://doi.org/10.1016/B978-0-12-819837-7.00003-0
17. Zakon o lekovima i medicinskim sredstvima. Sl. glasnik RS, br. 30/2010, 107/2012, 113/2017 - dr. zakon i 105/2017, Belgrade, Serbia (2017)
18. Zakon o medicinskim sredstvima. In: Sl. glasnik RS, br. 105/2017, Belgrade, Serbia (2017)
19. Melanie S, Roland H, Till B, Peter W, Bernd N, Patrick W, Claudia S (2020) Cotton and surgical masks-what ecological factors are relevant for their sustainability? Sustainability 12:10245. https://doi.org/10.3390/su122410245

Author Index

S. Kentera et al. (eds.), *Building Resilience in a Dynamic Critical Infrastructure Landscape*, NATO Science for Peace and Security Series C: Environmental Security,
https://doi.org/10.1007/978-3-032-04383-2

Index

S. Kentera et al. (eds.), *Building Resilience in a Dynamic Critical Infrastructure Landscape*, NATO Science for Peace and Security Series C: Environmental Security,
https://doi.org/10.1007/978-3-032-04383-2

MIX
Papier aus verantwortungsvollen Quellen
Paper from responsible sources
FSC® C105338
FSC
www.fsc.org

If you have any concerns about our products,
you can contact us on
ProductSafety@springernature.com

In case Publisher is established outside the EU,
the EU authorized representative is:
Springer Nature Customer Service Center GmbH
Europaplatz 3, 69115 Heidelberg, Germany

Printed by Libri Plureos GmbH
in Hamburg, Germany